Food Hydrocolloids

Volume III

Editor

Martin Glicksman
General Foods Corporation
Tarrytown, New York

CRC Press, Inc.
Boca Raton, Florida

Library of Congress Cataloging-in-Publication Data
(Revised for vol. 3)
Main entry under title:

Food hydrocolloids.

  Includes bibliographical references and indexes.
  1. Colloids. I. Glicksman, Martin.
TP453.C65F67 1982   664'.01   81-10018
ISBN 0-8493-6041-2 (v. 1)
ISBN 0-8493-6042-0 (v. 2)
ISBN 0-8493-6043-9 (v. 3)

Direct all inquiries to CRC Press, Inc., 2000 Corporate Blvd., N.W., Boca Raton, Florida, 33431.

© 1986 by CRC Press, Inc.
International Standard Book Number 0-8493-6041-2 (Volume I)
International Standard Book Number 0-8493-6042-0 (Volume II)
International Standard Book Number 0-8493-6043-9 (Volume III)

Library of Congress Card Number 81-10018
Printed in the United States

PREFACE

Volume III of this series of books on food gums and hydrocolloids continues with a pragmatic coverage of three important categories of gums, i.e., the cellulose gums, the plant seed gums, and the pectins. The chemical, physical, and functional properties of each of the important food gums in these categories are reviewed and discussed in relation with their utility in food product applications.

I would like to thank each of the contributing authors for their participation and assistance in this undertaking, and I also wish to express my gratitude to my many friends and associates in the food industry and at General Foods Corporation for their helpful advice, counsel, and encouragement. I specially want to thank Ms. Elly Cohen for her assistance with literature references and helpful library assistance and my grateful thanks to Mrs. Lillian Varian and Ms. Pat Lyon for typing and correcting the sections of the manuscript that I wrote.

Martin Glicksman
Valley Cottage, N.Y.
May 15, 1984

THE EDITOR

Martin Glicksman is a Principal Scientist in the Central Research Department of General Foods Corporation, Tarrytown, New York. For the past 30 years he has been actively involved in applied research and in the development of many new food products. Before joining General Foods he worked for seven years in the pharamaceutical and fine chemical industry as an organic chemist.

Mr. Glicksman has acquired an international reputation in the field of hydrocolloid technology, has published many papers, and holds 22 patents in the field. His best-known publication is the book *Gum Technology in the Food Industry,* which is a basic reference book on hydrocolloids in the food processing industry.

Mr. Glicksman holds a B.S. degree from City College of New York and M.S. and M.A. degrees from New York University. He is a Fellow of the Institute of Food Technologists and is currently a Counselor of the New York Section of the I.F.T. which has a membership of about 1100 food industry members. He is also on the Executive Board of the Carbohydrate Division of the national I.F.T. and serves on the editorial boards of the *Journal of Food Science* and *Carbohydrate Polymers.*

DEDICATION

To the memory of my friend,
Dr. (formerly Private) Martin Gold,
Comrade-in-arms at
Fort Benning, Georgia, and the
86th (Black Hawk) Infantry Division

CONTRIBUTORS

Steen Højgaard Christensen, M.Sc.
Manager, Application Research
Research and Development
A/S Københavns Pektinfabrik
Lille Skensved, Denmark

Martin Glicksman, M.S., M.A.
Principal Scientist
Central Research Department
General Foods Corp.
Tarrytown, New York

Joseph A. Grover, Ph.D.
Research Associate
Michigan Applied Science and
 Technology Laboratories
The Dow Chemical Co.
Midland, Michigan

Carl T. Herald, Ph.D.
Product Sales Manager
Celanese Water Soluble Polymers
Louisville, Kentucky

John D. Keller, M.S.*
Manager, Research and Development
Food Hydrocolloid Group
Hercules, Inc.
Middletown, New York

**Christopher McIntyre, L.R.S.C.,
 A.I.F.S.T.**
Director, Technical Services
PFW Division
Hercules, Inc.
Food Technology Center
Middletown, New York

William R. Thomas
Product Marketing Manager
Marine Colloids Division
FMC Corp.
Philadelphia, Pennsylvania

*   Now with Nestlé S. A., New Milford, Connecticut.

FOOD HYDROCOLLOIDS

Volume I

Volume II

Volume III

TABLE OF CONTENTS

I. CELLULOSE GUMS

II. PLANT SEED GUMS

III. PLANT EXTRACTS

Cellulose Gums

INTRODUCTION

Martin Glicksman

TABLE OF CONTENTS

I. BACKGROUND

Cellulose — the backbone of the vegetable world — is the most plentiful renewable resource in the world. It has been estimated that between 10^{10} and 10^{11} tons are synthesized and destroyed each year.[8] Cellulose thus is an extremely good economic starting material for the development of new functional polymers.

Cellulose is the major constituent of all vegetation, comprising from one third to one half of dry-plant material. Cellulose, together with hemicelluloses and lignins, are the major structural constituents of most land plants and form the cell walls and intercellular layers of the basic supporting structure of the plant. About 40 to 50% of all woody materials and up to 98% of cotton linters are made up of cellulose. Other agricultural residues such as corn stalks, corn cobs, and wheat straw contain smaller amounts of cellulose (about 30%), but are all available as a giant reservoir of potentially available cellulosic raw material.[1,2]

II. STRUCTURE

Cellulose is a linear polymer of D-glucose monomers joined by β-D-(1→4) - linkages and arranged in repeating units of cellobiose, each composed of two anhydroglucose units. It has a high molecular chain length, and the hydrogen-bonding capacity of the three hydroxyl groups is very great. Because of the uniform nature and linear structure of the molecules, they can fit together over at least part of their length to form crystalline regions of great strength and rigidity. These crystalline regions give plant cell walls their great strength and rigidity as exemplified by the high tensile strengths of all vegetable fibers.[1,2,7,10]

In nature, cellulose occurs in the presence of other polysaccharides such as xylan and mannan, which are often known as "hemicelluloses". Cellulose is purified by removal of these materials, as well as proteins and lipids, etc. by boiling in dilute alkali.

The degree of polymerization (DP) of cellulose ranges from 3900 to 18,000,[8] and each D-glucose component is in the pyranose form in the C1(D) chair conformation, the complete polymer being arranged spatially into long thread-like molecules, as confirmed by X-ray measurements of native cellulose. The cellulose molecules are aligned to form fibers, some regions of which are highly ordered and have a crystalline structure due to lateral association by hydrogen bonding. The crystalline regions vary in size and represent areas of great mechanical strength and high resistance to attack by chemical reagents and hydrolytic enzymes. The physical and chemical properties of cellulose are greatly dependent on the relative amount and arrangement of the crystalline regions. The cellulose molecules tend to remain extended but may normally undergo a degree of turning and twisting. Because of its size and strong associative forces, it can only be brought into solution under certain conditions, usually by chemically modifying the polymer to add side chains that will separate the polymers and allow hydration and solubilization in aqueous media to take place. In this way, many diverse and useful functional properties can be imparted to the cellulose molecule.[1,2,7,9,10]

The cellulose derivatives most useful in the food industry are ethers in which alkyl or hydroxyalkyl groups have been substituted upon one or more of the three available hydroxyl groups in each anhydroglucose unit of the cellulose chain. The effect of the substituent groups is to disorder and spread apart the cellulose chains so that water or other solvents may enter to solvate the chain. By controlling the type and degree of substitution, it is possible to produce products that have a wide range of functional properties. The more important water-soluble derivatives are those with the following substituent groups:

NaOOC—CH$_2$—	Sodium carboxymethyl
CH$_3$—	Methyl
HO—CH$_2$—CH$_2$—CH$_2$— or CH$_3$—HOCH—CH$_2$—	Hydroxypropyl
CH$_3$—HOCH—CH$_2$— and CH$_3$—	Methylhydroxypropyl

When all three available hydroxyl positions on the cellulose molecule are replaced by a substituent group, the derivative is said to have a degree of substitution (DS) of three. Actually, this is rarely the case, since partial substitutions are preferred, but it can readily be seen that varying the degree of substitution as well as the type of chemical substituent makes possible a tremendous range of permutations and combinations offering a wide selection of functional properties. However, only a comparatively limited number of these derivatives have been found to have industrial or food applications.[1,2,10]

III. PREPARATION

All D-glucose units in the cellulose polymer chain have hydroxyl groups available at carbons C-2, C-3, and C-6 which can be substituted by etherification or esterification (Figure 1).

When all the hydroxyl groups are substituted, the cellulose is said to have its maximum degree of substitution of 3.0 (DS of 3.0). Most commercial water-soluble cellulose derivatives have substantially less substitution, usually in the area of DS 1.0. When side chain formation is possible, the substituent groups are defined in molar terms and the molar substitution (MS) value can exceed 3.0.

The properties of a specific cellulose ether depend on the type, distribution, and uniformity of the substituent groups. The water-soluble cellulose ethers possess a range of multifunctional properties resulting in a broad spectrum of end uses and applications.

The preparation of water-soluble derivatives, mostly ethers, essentially begins with the same initial step — treatment of the cellulose with aqueous sodium hydroxide to swell and distend the fibers. This is followed by reaction with the appropriate reagents to form the desired compounds.

The water-soluble cellulose ethers important to the food industry are prepared in the following ways:

1. **Carboxymethylcellulose (CMC)**[5,12] — CMC is produced by treating cellulose with aqueous sodium hydroxide followed by reaction with sodium chloroacetate:

$$R—OH + NaOH + ClCH_2COONa \rightarrow R—OCH_2COONa + NaCl + H_2O$$

A side reaction, the formation of sodium glycolate, also occurs:

$$ClCH_2COONa + NaOH \rightarrow HOCH_2COONa + NaCl$$

2. **Methylcellulose and Hydroxypropylmethylcellulose**[4,11] — Methylcellulose is prepared by reacting purified wood pulp or cotton linters having a high α-cellulose content with aqueous sodium hydroxide, and then with methyl chloride according to the following reactions where R is the cellulose radical:

●Main reactions:

$$R-OH + NaOH \rightarrow R-OH \cdot NaOH \text{ (complex)}$$
$$R-OH \cdot NaOH \rightleftarrows R-ONa + H_2O$$
$$R-ONa + CH_3Cl \rightarrow R-OCH_3 + NaCl$$

●Side reactions:

$$CH_3Cl + NaOH \rightarrow CH_3OH + NaCl$$
$$CH_3Cl + H_2O \rightarrow CH_3OH + HCl$$
$$CH_3OH + NaOH \rightleftarrows CH_3ONa + H_2O$$
$$CH_3ONa + CH_3Cl \rightarrow CH_3OCH_3 + NaCl$$

For the production of hydroxypropylmethylcellulose, propylene oxide is also added to the mixture and reacts as follows:

●Main reaction:

$$R-OH + CH_2\overset{O}{\overbrace{\quad}}CH \underset{CH_3}{\bigg|} \xrightarrow{NaOH} R-O-CH_2-\underset{\underset{\displaystyle CH-CH_3}{}}{\overset{OH}{\overset{|}{}}}$$

●Side reaction:

$$CH_2\overset{O}{\overbrace{\quad}}CH-CH_3 \xrightarrow[CH_3Cl]{NaOH} \text{glycols + glycol ethers}$$

The relative amounts of methyl and hydroxypropyl substitution are controlled by the weight ratio and concentration of sodium hydroxide and the weight ratios of methyl chloride and propylene oxide per unit weight of cellulose.

3.   **Hydroxypropylcellulose**[6,13] — Cellulose derived from wood pulp or cotton linters is treated with aqueous sodium hydroxide, and the resulting alkali cellulose is reacted with propylene oxide:

$$R-OH + CH_3\overset{O}{\overset{/\backslash}{}}CHCH_2 \rightarrow R-OCH_2CHOHCH_3$$

The secondary hydroxyl group in the hydroxypropyl group is capable of hydroxypropylation to give a side chain:

$$R-OCH_2CHOHCH_3 + CH_3\overset{O}{\overset{/\backslash}{}}CHCH_2 \rightarrow R-\underset{\underset{\displaystyle CHCHCH_3}{}}{\overset{CH_2CHOHCH_3}{\overset{|}{\underset{O}{\overset{|}{}}}}}$$

The sodium hydroxide functions as a swelling agent and a catalyst for the etherification. A side reaction in which propylene oxide reacts with water to form a mixture of

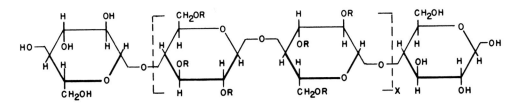

FIGURE 1.   Structure of cellulose and derivatives.

propylene glycols also occurs, but can be minimized by keeping the water input as low as possible.

IV. PROPERTIES

Of the many cellulose polymers and derivatives investigated and manufactured, these four water-soluble polymers have found utility in the food industry. In addition, pure cellulose in specially prepared, physically modified form[9] has been shown to have useful functional hydrocolloidal properties and has found significant use in certain food applications. These five materials will be discussed in the four chapters of this section of the book.

Sodium carboxymethylcellulose (cellulose gum or CMC) is the most important cellulose-derived hydrocolloid used in the food industry. It is an anionic polymer, and in addition to its ability to thicken and modify rheology of water solutions, it has the unique property of being able to react with proteins and other charged molecules within specific pH ranges. These functional properties have led to its extensive use in the food industry in many diverse product applications.[3,5,12]

Methylcellulose (MC) and hydroxypropylmethylcellulose (HPMC) are polymers having the useful properties of thickening, surfactancy, film forming ability, adhesiveness, and the unique property of thermal gelation. They have the unusual property of solubility in cold water and insolubility in hot water, so that when a solution is heated, a three-dimensional gel structure forms, which is reversible when the solution cools. These properties have led to widespread industrial applications and limited, specialized applications in the food industry.[4,11]

Hydroxypropylcellulose (HPC), sold under the trade name of Klucel®,[6] is a nonionic cellulose ether with an unusual combination of properties. These include solubility in both water (below 40°C) and polar organic solvents, high surface activity, and thermoplasticity plastic flow (extrudability under pressure) properties; and the insensitivity of viscosity to changes in pH.[6,13]

Microcrystalline cellulose (MCC), sold under the trade name of Avicel®, is really a family of materials having hydrocolloidal functional properties which has resulted in their use as hydrocolloid ingredients in food applications. The basic microcrystalline cellulose is an acid-hydrolyzed, pure α-cellulose material that has effective thickening and water-absorptive properties when dispersed under high shear conditions. To improve hydration rate and capacity, several of the microcrystalline cellulose materials have to be dried with other materials such as CMC. This has improved their functionality and extended their utility in food products application, especially in the area of dairy and frozen food products.[1,9]

REFERENCES

1. **Glicksman, M.,** *Gum Technology in the Food Industry,* Academic Press, New York, 1969.
2. **Whistler, R. L.,** *Industrial Gums,* 2nd ed., Academic Press, New York, 1973.
3. **Ganz, A. J.,** Cellulose hydrocolloids, in *Food Colloids,* Graham, H., Ed., AVI Press, Westport, Conn., 1977, 382—417.
4. Dow Chemical Co., *Handbook on Methocel® Cellulose Ether Products,* Dow Chemical Co., Midland, Mich., 1981.
5. Hercules, Inc., *Cellulose Gum — Chemical and Physical Properties,* Hercules, Inc., Wilmington, Del., 1978.
6. Hercules, Inc., *Klucel® Hydroxypropyl Cellulose — Chemical and Physical Properties,* Hercules, Inc., Wilmington, Del., 1976.
7. **Ward, K., Jr. and Seib, P. A.,** Cellulose, lichenan and chitin, in *The Carbohydrates — Chemistry and Biochemistry, Vol. 2A, 2nd ed.,* Pigman, W., Horton, D., and Herp, A., Eds., Academic Press, New York, 1970, 413—445.
8. **Walton, A. G. and Blackwell, J.,** *Biopolymers,* Academic Press, N.Y., 1973, 464—474.
9. **Battista, O. A.,** *Microcrystal Polymer Science,* McGraw-Hill, New York, 1975.
10. **Whistler, R. L. and Zysk, J. R.,** Carbohydrates, in *Encyclopedia of Chemical Technology,* Vol 4, 3rd ed., Grayson, M. and Eckroth, D., Eds., John Wiley & Sons, New York, 1978, 535—555.
11. **Greminger, G. K., Jr. and Krumel, K.,** Alkyl and hydroxyalkylalkylcellulose, in *Handbook of Water-Soluble Gums and Resins,* Davidson, R. L., Ed., McGraw-Hill, New York, 1980, chap. 3.
12. **Stelzer, G. I. and Klug, E. D.,** Carboxymethylcellulose, *Handbook of Water-Soluble Gums and Resins,* Davidson, R. L., Ed., McGraw-Hill, New York, 1980, chap. 4.
13. **Butler, R. W. and Klug, E. D.,** Hydroxypropylcellulose, in *Handbook of Water-Soluble Gums and Resins,* Davidson, R. L., Ed., McGraw, Hill, New York, 1980, chap. 13.

Chapter 1

MICROCRYSTALLINE CELLULOSE (MCC OR CELLULOSE GEL)

William R. Thomas

TABLE OF CONTENTS

I. INTRODUCTION

Microcrystalline cellulose is one of the few relatively new ingredients available to the food industry. The first commercial quantities were produced in 1962; however, MCC did not become a factor in food stabilization until very late in the 1960s. Although it was first believed that MCC was the ideal "low calorie" ingredient for foods, there has been only limited success in that area. By far, the major uses of edible microcrystalline cellulose are for pharmaceutical tableting and the functional (stabilizing) applications for foods.

Microcrystalline cellulose results from the concentration of the naturally occurring crystalline portion of the cellulose fiber after acid hydrolysis to the level of DP. It is not a chemical derivative, but a purified native cellulose reduced from a fibrous state to a crystalline powder or a redispersible colloidal gel.

The evaluation and application of microcrystalline cellulose (cellulose gel) in food products must take into consideration the science of fine particle technology and be differentiated from the technology associated with starch and soluble gums.

II. PRODUCTION

To understand the application of microcrystalline cellulose, it is necessary to understand its source, manufacturing, and how it differs from other hydrocolloids.

Cellulose is one of the most abundant and renewable natural resources available to mankind. Cellulose is a polysaccharide polymer composed of glucose units, linked in a linear B14 configuration, which forms the fibrous tissue of plant life. Therefore, it is present in much of the food we eat.

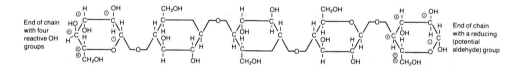

The cellulose polymer is thread-like and studded with hydroxyl groups. These hydroxyl groups are responsible for very strong lateral forces between adjacent molecules which hold together bundles of chains into these fibrils. The molecules in the fibril bundle are arranged in sufficient regularity to defract X-rays in a characteristic pattern in the same way that inorganic crystals do. Microfibrils are then arranged in layers to form the cell wall of plant fibers.

During the manufacture of MCC, the dissolving alpha cellulose pulp is treated with a dilute mineral acid and the cellulose microfibril is unhinged. This hydrolysis process is carried out until a point of leveling off of degree of polymerization is obtained. The para-crystalline regions (Figure 1), areas of disordered molecular structure, are weakened as the acid selectivity etches away this area from around the densely arranged crystalline cellulose, similar to removing the mortar from around bricks (Figure 2). Subsequent shear releases the cellulose aggregates, which are the basic raw materials for commercial microcrystalline cellulose products.

Cellulose Microfibrils

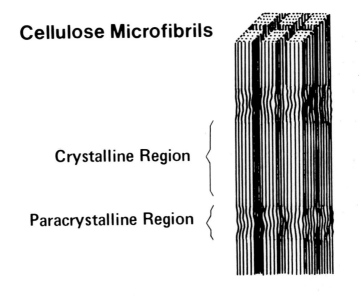

Crystalline Region {

Paracrystalline Region {

FIGURE 1. Cellulose microfibrils.

The manufacturing schematic shown in Figure 3 is helpful in differentiating the major types of cellulose products available in the food industry.

Soluble gums, such as CMC, are produced by first treating or opening up the molecular structure of alpha cellulose so that it can be chemically derived. In comparison, microcrystalline cellulose remains insoluble and is not derived during the acid treatment and drying steps. Hence, it remains native cellulose.

A second major product from cellulose is fibrous floc which is basically pulverized or wet-milled alpha cellulose.

A. Powdered Microcrystalline Cellulose

The manufacturing process most common to microcrystalline cellulose produced two divergent products. The first type of product is powdered microcrystalline cellulose (Figure 4). This powdered MCC is a spray-dried MCC aggregate that ranges from 2 to 200 μg and is porous, plastic, and sponge-like. This powdered nonfibrous form of microcrystalline cellulose is used for tableting, dietary applications, as a flavor carrier, and as an anticaking agent for shredded cheese.

B. Colloidal Microcrystalline Cellulose

The second major type of microcrystalline cellulose is water-dispersible colloidal cellulose that has similar function as soluble gums.

Major Functions of MCC

Powdered MCC	Colloidal MCC
Binder/disintegrant for tablets	Emulsion stabilizer
Flow aid for cheese	Thixotropic thickener
Carrier for flavors	Moisture control
Fiber	Foam stabilizer
	Provide cling
	Ice crystal control
	Suspending agent

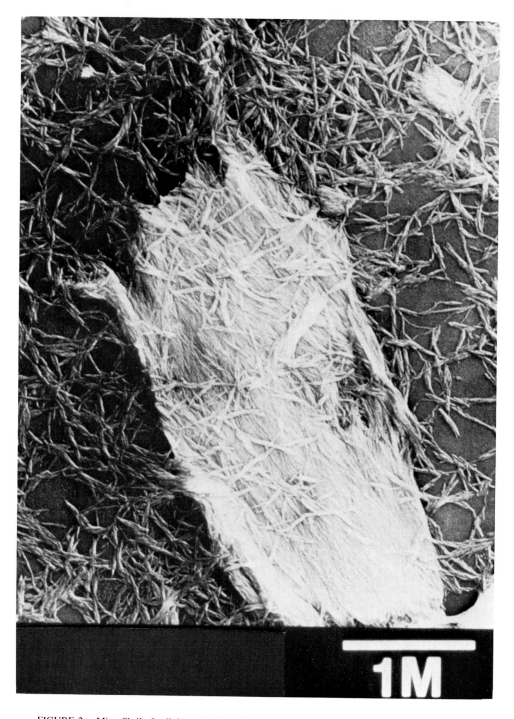

FIGURE 2.   Microfibril of cellulose releasing microcrystals of cellulose. (Magnification × 50,000.)

In the manufacturing of colloidal types, considerable energy in the form of mechanical attrition is put into the cellulose pulp after hydrolysis, in order to tear the weakened microfibril apart and complete the unhinging process. Current production techniques provide 90% yields of 60 and 70% colloidal crystallite aggregates below 0.2 μm.

Cellulosic Products

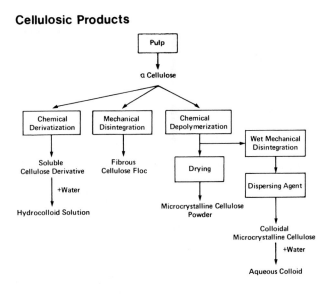

FIGURE 3. Manufacturing process for major types of cellulose food ingredients.

If the colloidal cellulose were dried at this point, hydrogen bonding would cause hornification of cellulose, preventing it from becoming functional as a hydrocolloid stabilizer. For this reason, CMC is used to act as a barrier during drying to prevent this hornification. The CMC also is a dispersant to aid in the redispersion of the insoluble microcrystals in aqueous systems. In most colloidal grades of microcrystalline cellulose (cellulose gel), 60 to 70% of these microcrystals will be below 0.2 μg and function as a stabilizer after redispersion in aqueous medium. These submicron particles do not hydrate, but disperse; hence, the use of minute particles to stabilize food systems. Although the microcrystalline cellulose itself forms a thixotropic gel, CMC adds additional stability and can be used to alter the rheology of the system.

III. TECHNICAL PROPERTIES

A. Powdered Microcrystalline Cellulose Types

The spray-dried agglomerated aggregate of microcrystalline cellulose is sold in various size ranges typified by these FMC Corporation grades:

Powdered Microcrystalline Cellulose Types

	Average particle size (μm)
PH-101	50
PH-102	90
PH-105	18

The different configurations and particle size ranges affect flow properties in tableting, oil and water absorptive capacity, as well as bulk density. Although the MCC is totally carbohydrate, it is not metabolized; therefore, it may be used in low-calorie foods or as a source of food fiber.

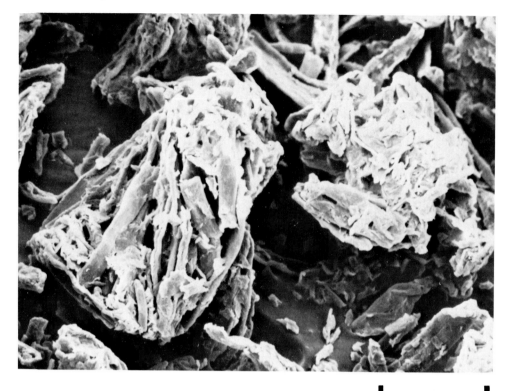

100 μm

FIGURE 4.   Avicel® MCC type PH.102.

Typical Properties of MCC

Appearance	White, odorless, tasteless, free-flowing powder
Heavy metals	<10 ppm
Water soluble substances	<8 mg/5 g
Residue on ignition	0.1%
pH (all except for PH-105)	5.5 go 7.0
pH (for PH-105)	5.0 to 7.0

B. Colloidal Microcrystalline Cellulose

For purposes of this book which discusses functional hydrocolloids, this is the micro-crystalline cellulose product which must be considered. The development of colloidal microcrystalline cellulose has provided the food formulator with a food grade submicron particle which can be used to stabilize food systems, similar to the way expanding lettuce clays are used for industrial purposes. Microcrystalline cellulose is a bundle of short chain cellulose units, which have an accumulative molecular weight of many millions. One must realize that in using microcrystalline cellulose as a stabilizer, we are dealing with particle and surface technology and not molecular science.

The product codes used by FMC Corporation are helpful in describing the commercially available products (Table 1). These products vary substantially in functionality, composition, raw materials source, and method of application; they are not interchangeable. For example, RC-501 and RC-581 must be homogenized, whereas CL-611 and RC-591 require only good

Table 1
COMMERCIAL COLLOIDAL MCC TYPES

	MCC (%)	FMC designated	Required equipment to activate	Use levels in foods (%)	Colloidal (%)	Set-up viscosity	Primary use
Bulk-dried	91	RC-501	Homogenizer	0.5 to 3	30	2.1% = 750—1300 cps	Whipped topping Heat stable emulsion
Bulk-dried	89	RC-581	Homogenizer	0.3 to 0.8	70	1.2% = 800—1450 cps	Frozen desserts Emulsions
Spray-dried	89	RC-591	High-speed mixer	0.3 to 1.0	70	1.2% = 900—1600 cps	General food use — thixotropic gel
Spray-dried	85	CL-611	High-speed mixer	0.3 to 2.5	70	1.2% = 50—300 cps	Pourable systems
Whey/MCC	22	WC-595	Slow mixing	2 to 4	15	4.0% = 500—2000 cps	Dry blended foods

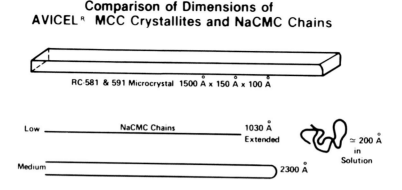

FIGURE 5.   Comparison of MCC and CMC.

agitation, but WC-595 may be activated and become functional in a food system by simply spoon stirring. RC-501 is 30% colloidal as compared to RC-581, which is 60 to 70% colloidal. The more colloidal the MCC, the smaller the amount required to stabilize. The larger the microcrystal is in size, the more white the end product. Depending on the percent colloidal microcrystalline cellulose and the CMC barrier used, products will vary in their thixotropy as well as their opacity (Figure 5).

Again, this colloidal material must be dispersed (peptized) in an aqueous medium to become functional. For example, a spray-dried agglomerate of microcrystalline cellulose can contain many, many millions of microcrystals as described in Figure 6.

This agglomerate must be peptized and the microcrystals released to become functional. MCC does not swell like starch or hydrate like a gum, but is dependent on shear and available water. Salts, acids, and other ingredients will also play a role in the peptization of colloidal microcrystalline cellulose. Heat has very little effect.

These submicron microcrystals, when dispersed into water, form suspensions and thixotropic gels after a certain concentration is reached, depending on the type of microcrystalline cellulose.

The stability of the system is further enhanced by the leashing of the insoluble microcrystals with CMC (Figure 7), which forms a barrier on drying. The rheology and the stability of the system can be further modified by additional protection colloids such as CMC or xanthan gum.

1. Flocculation by Electrolytes

Dispersions of colloidal MCC are flocculated by small amounts of electrolyte, cationic polymers, and surfactants, whereas the nonionic soluble cellulose derivatives have higher resistance to flocculation. MCC dispersions are compatible with most nonionic and anionic polymers. CMC, hydroxypropylmethylcellulose, and xanthan gum have been found to be the best protective colloids for MCC dispersions. Ultimately, the flocculation value and compatibility of RC-591/CMC and RC-591/HPMC blends become the same as solutions of CMC and HPMC.

The flocculation values for most colloidal MCCs are approximately the same as those for Avicel® RC-501 listed in Table 2.

When xanthan gum is used as a protective colloid, a ratio of ten parts RC-591 to one part xanthan gum is recommended.

Most charts such as Table 3 are based on dilute systems which have little bearing on food products, except beverages. A primary example of this interpretation is the flocculation of MCC in dilute systems by galactomannans. In reality, these food systems are a combination of fats, starches, sugars, etc. and the phenomena is unimportant. The single most important

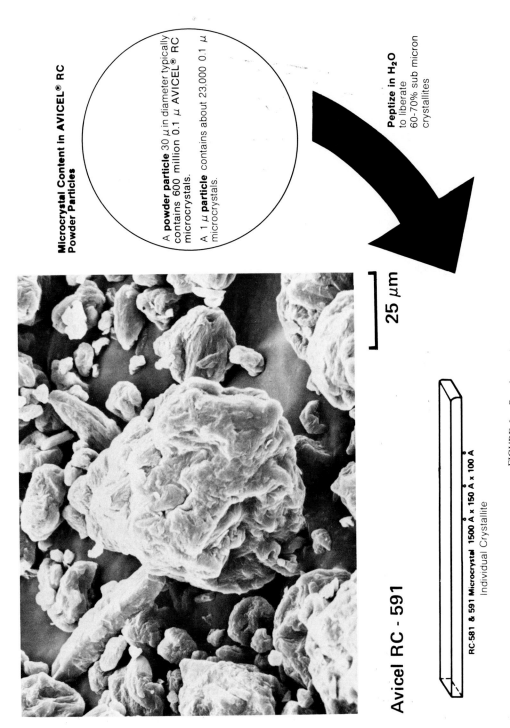

Microcrystal Content in AVICEL® RC Powder Particles

A **powder particle** 30 μ in diameter typically contains 600 million 0.1 μ AVICEL® RC microcrystals.

A 1 μ **particle** contains about 23,000 0.1 μ microcrystals.

Peptize in H₂O to liberate 60-70% sub micron crystallites

Avicel RC - 591

25 μm

RC-581 & 591 Microcrystal 1500 Å x 150 Å x 100 Å
Individual Crystallite

FIGURE 6.   Powder particle of MCC.

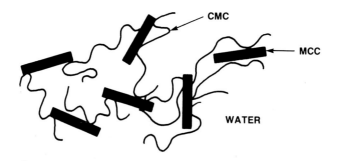

FIGURE 7.   Microcrystals are leashed together with soluble polymers.

thing in the application of microcrystalline cellulose is the peptization of the agglomerates. Principal factors affecting this peptization are as follows:

- Type of microcrystalline cellulose
- Water hardness
- Amount of shear available
- Order of addition of ingredients

2. Quality Control

The type of microcrystalline cellulose selected will be directly influenced by the hardness of the water and the type of dispersion (shear) equipment available. This interrelationship is apparent by reviewing Figure 13. The harder the water and the more di- and trivalent electrolytes in the water for formulation, the more shear required. If the shear is not available, then a colloid MCC type which is spray-dried must be selected.

Trouble-free predictable food systems can be easily developed and controlled by following these few simple rules:

- Select the MCC product based on the needed function, water hardness, shear available, price, etc.
- Microcrystalline cellulose is to be the first ingredient added to the water.
- Agitate for 10 min.
- Add salts and acids last.
- Apply shear required, depending on microcrystalline cellulose selection and food product requirements.

The necessity for MCC being one of the first ingredients added to the food system is threefold. First, if the MCC becomes coated with fat, then the water cannot swell the agglomerate and release the microcrystals. Second, MCC requires considerable free water and cannot compete with other gums or starches for that water. The fewer electrolytes in the system prior to the activating of microcrystalline cellulose, the more effective and the higher the percent of colloidal material released.

The best way to perform QC checks on the activated MCC is to view it under polarized light. Figure 8 shows the microcrystalline cellulose unpeptized, fully peptized, and flocculated. In most applications, full peptization of the agglomerate is desirable, which means less microcrystalline cellulose will be required and the end product will be more consistent.

3. Selection of Colloidal Microcrystalline Cellulose (Cellulose Gel) Types — Comparison of Colloidal MCC with Powdered Type MCC

The photograph in Figure 9 illustrates the dramatic differences in the dispersion charac-

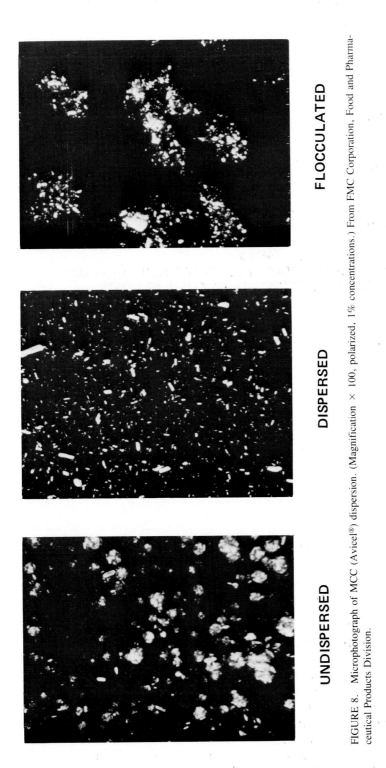

FIGURE 8. Microphotograph of MCC (Avicel®) dispersion. (Magnification × 100, polarized, 1% concentrations.) From FMC Corporation. Food and Pharmaceutical Products Division.

Table 2

**DISPERSION OF AVICEL® RC-581 AND AVICEL® RC-591 IN
DISTILLED WATER AND CITY WATER**

		Mixing devices			
Product	Water	Propeller mixer low speed $^1/_2$ hr	Propeller mixer high speed $^1/_2$ hr	Centrifugal pump recirculating $^1/_2$ hr	Waring Blendor® 2 min 100 V
RC-581	Distilled	34.8%[a]	74.7%	73.2%	100.0%
RC-591	Distilled	100.0	100.0	100.0	100.0
RC-581	City — 264 umhos	4.0	6.6	33.9	82.7
RC-591	City — 264 umhos	12.6	22.2	100.0	100.0

[a]   Completeness of dispersion (peptization) is expressed in %; 100% = complete peptization, 0% = no peptization.

Table 3

**NaCl and AlCl₃ FLOCCULATION VALUES FOR AVICEL® RC-501
DISPERSIONS PROTECTED BY CMC AND MC 90 HG**

Sample composition	Avicel® RC/ soluble polymer solids ratio[a]	Flocculation value[b]	
		NaCl (N)	AlCl₃(N)
Avicel® RC/CMC-7MP	100/0	10^{-2}—10^{-2}	10^{-4}—10^{-3}
	90/0	2—3	10^{-4}—10^{-3}
	80/20	2—3	10^{-4}—10^{-3}
	70/30	3—4	10^{-4}—10^{-3}
	50/50	S	10^{-4}—10^{-3}
	0/100	S	10^{-4}—10^{-3}
Avicel® RC/Methocel® 90 HG, 4000 cps	100/0	10^{-2}—10^{-1}	10^{-4}—10^{-3}
	90/10	0.5—1	10^{-4}—10^{-3}
	80/20	1—1.5	10^{-3}—10^{-2}
	70/30	1.5—2	10^{-2}—10^{-1}
	50/50	1.5—2	S
	0/100	1.5—2	S

[a]   The numerator is Avicel® RC-501 (91.5 weight parts of Avicel® microcrystalline cellulose and 8.5 parts of sodium carboxymethylcellulose). The denominator represents additional soluble polymer.

[b]   The flocculation value is the concentration of electrolyte required to flocculate a 0.2% solids dispersion. Evaluation made 2 days after make up. S denotes stable in saturated salt solution.

teristics of MCC products. The 0.5% microcrystalline cellulose (cellulose gel) in each of the three cylinders has been agitated in distilled water with a magnetic laboratory stirrer for 5 min and then allowed to stand undisturbed for 24 hr.

Cylinder 1 on the left shows little activation (peptization) of the powdered MCC which it contains. This was the first dry microcrystalline cellulose developed and will not form a colloidal gel. It is used extensively by the pharmaceutical industry for tableting and in the food industry as an anticaking agent for cheese.

Cylinder 2 shows a cloudy dispersion which contains the bulk-dried microcrystalline cellulose. This type of microcrystalline cellulose requires high shear equipment — specifically homogenizers — to activate the microcrystalline cellulose.

Cylinder 3, which contains spray-dried microcrystalline cellulose, is completely activated

FIGURE 9.   Visual comparison of dispersion characteristics of MCC PH-101, RC-581, and RC-591.

or peptized. In such a dispersion, approximately 70% of the particles are colloidal and less than 0.2 μm.

4. Rheological Properties

Microcrystalline cellulose gels are highly thixotropic and have a finite yield value at low concentrations, as shown in Figure 10. The yield value and elasticity are due to the linking of the solid particles to form a network. When the gel is sheared, the particles separate and act independently, breaking the network and the gel thins as a result of that shear. When the dispersion is allowed to rest after shearing, a yield value is reestablished quickly, within minutes, and the solid particles reform into a network.

When CMC is added to colloidal gels of MCC, the soluble polymer molecules absorb onto the MCC particles. As the proportion of the CMC is increased, the gels become less thixotropic and their yield value decreases, as exemplified by the 50/50 mixture shown in Figure 10. At the same time, the resistance to flocculation increases. A similar behavior is shown by mixtures of microcrystalline cellulose products with methylcellulose.

The rheological properties of microcrystalline cellulose colloidal gels are modified by the addition of carboxymethylcellulose (CMC), methylcellulose (MC), xanthan gum, and other various gums. The effect of CMC on rheological behavior is shown in Figure 11.

Changes in temperature have little effect on the viscosity of microcrystalline cellulose colloidal dispersions. Figure 12 shows the reversible change of the viscosity in a colloidal MCC dispersion and a CMC solution over a temperature range from 25 to 80°C. Additional data has indicated that MCC in a sealed container at 120°C for 75 min will not alter the physical properties of MCC colloidal gels.

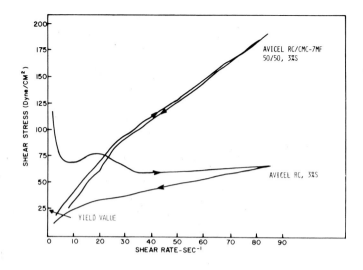

FIGURE 10.   The flow properties of an MCC dispersion in a 50/50 blend of
MCC/CMC-7MF.

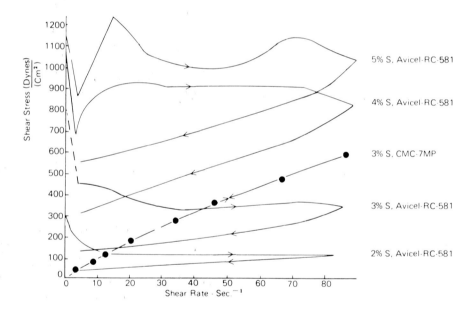

FIGURE 11.   Shear stress as a function of increasing and decreasing shear rate of MCC gels.

5. Comparison of Dispersibility of Bulk-Dried Colloidal MCC and Spray-Dried Colloidal MCC in Distilled and City Water

Table 2 shows the dramatic improvement in the activation (peptization) characteristics of spray-dried RC-591 over bulk-dried RC-581. It also illustrates the effect that different mixing devices have on the dispersion of these two microcrystalline cellulose products in both distilled and city water. When peptizing MCC, it is important to know the water hardness and the type of electrolytes that the water contains. In distilled water, microcrystalline cellulose type RC-591 can be peptized 100% with almost any mixing device. But in city water of average hardness (265 umhos), it requires a recirculating centrifugal pump, colloidal mill, or homogenizer for complete activation. However, spray-dried type RC-591 can always be dispersed easily if one selects the proper mixing device for the type of water used. It

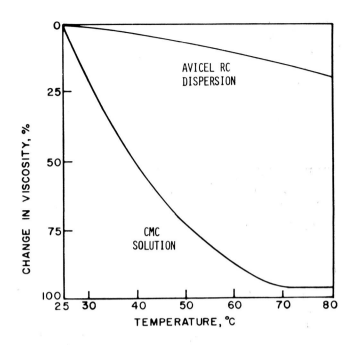

FIGURE 12. Change in viscosity.

should be noted at this point that in applications other than cloud suspensions for beverage applications, 100% dispersion of the microcrystalline cellulose is not required.

In food formulations, salt and acids can be a problem unless they are added last.

Recent developments resulting in new products, which are combinations of sprayed microcrystalline cellulose and whey, now allow for the use of microcrystalline cellulose in dry blended foods. In these products, the microcrystalline cellulose is preactivated and then dried in a 3:1 ratio of whey to microcrystalline cellulose; water hardness is no problem whatsoever.

6. Recommended Equipment for Dispersing Spray-Dried Microcrystalline Cellulose in Monovalent and Divalent Salt Solutions

As illustrated in Figure 13, the spray-dried MCC is relatively easy to disperse (peptize) in a monovalent salt solution. All that is needed at the most common concentration levels of these salts is a high-speed propeller mixer to provide adequate shear. However, Figure 13 also shows that to disperse spray-dried microcrystalline cellulose in divalent salt solutions, a centrifugal pump is needed since higher shear is required. Although Figure 13 does not take into consideration specific salts or combinations of salts, they do serve as an excellent guide in selecting dispersion equipment. Of course, if any further information or assistance is needed, it is always available from the suppliers.

Spray-dried microcrystalline cellulose (cellulose gels) can be dispersed in almost any condition with a colloidal mill or homogenizer, but when working with food formulations three rules should be adhered to when possible:

1. Disperse microcrystalline cellulose in the water before adding any other ingredients to avoid competition for available water. Agitate for 10 min.
2. Add all other ingredients before lowering pH or adding salt.
3. Remember when using bulk-dried microcrystalline cellulose products such as RC-581 and -501, they must be homogenized.

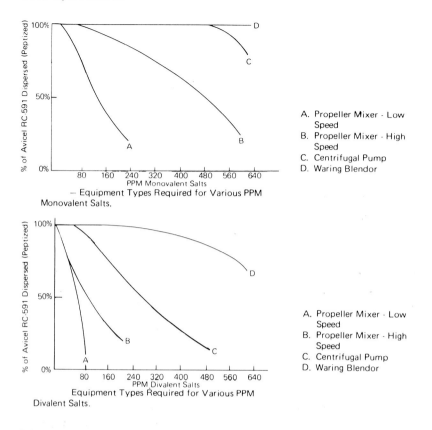

FIGURE 13.   Shows the dispersion of spray-dried microcrystalline cellulose in monovalent and divalent salt solutions using various mixing devices.

Table 4
DISPERSION OF AVICEL® RC-591 USING SEQUESTRANTS

	Specific conductance (umhos)	Propeller mixer (¹/₂ hr)	
		1000 rpm low speed	2500 rpm high speed
Distilled + 100 ppm $CaCl_2$	222.6	6.2%	32.3%
Distilled + 100 ppm $CaCl_2$, 150 ppm sodium citrate	259.8	13.9%	75.7%

Note: Completeness of dispersion is expressed in %.

7. Use of Sequestrants to Improve Dispersibility of Spray-Dried Microcrystalline Cellulose in Water with High Electrolyte Content

As shown in Table 4, sodium citrate definitely improves the dispersibility of spray-dried microcrystalline cellulose products in a $CaCl_2$ solution. These results and other research work indicate that certain sequestrants can be of help where dispersion problems are encountered in hard water supplies.

8. Compatibility with Other Hydrocolloids

Colloidal microcrystalline cellulose (cellulose gel) is compatible with most hydrocolloids

Table 5
COMPATIBILITY OF
AVICEL® RC-591 WITH
MISCIBLE LIQUIDS

Parts of 1% Avicel® RC-591
dispersion/parts of water-
miscible liquid

50/50 Acetone
40/60 Ethanol
40/60 Glycerine
50/50 Methanol
40/60 Propylene glycol
40/60 Sorbitol solution (70%)

in use today; however, it is generally conceded that both guar gum and sodium alginate will slightly inhibit the functionality.

9. Compatibility with Miscible Liquids

Colloidal MCC (cellulose gel) blends will not form colloidal dispersions in alcohols or glycols. However, they do have an excellent tolerance for these liquids and for many other water-miscible liquids (Table 5). Once formed, an aqueous dispersion can have large amounts of numerous water-miscible liquids added without causing flocculation.

IV. APPLICATION IN FOODS

A. How Colloidal Microcrystalline Cellulose (Cellulose Gel) Functions

To be functional in a food system, the dried agglomerant of crystallites must be peptized, so that individual microcrystals are released into the aqueous phase of the food. When sufficient number of microcrystals are dispersed into the aqueous phase, stabilization occurs. Remember, these crystallites do not hydrate but merely disperse (peptize) into the aqueous phase.

The dispersion (peptization) of insoluble 0.2-μm cellulose particles is aided by the CMC which is used in the manufacturing to prevent hydrogen bonding and to improve the taste. The rapid hydration of the CMC not only propels the microcrystals from their dried matrix, but also links the crystallites together after rehydration to form a thixotropic gel network. This thixotropic gel is very stable with heat, shear, and pHs down to 3.5. Its stability to high salt levels and at lower pHs can be dramatically increased by the addition of protective colloids, like xanthan gum and CMC.

As noted in Figure 14, the microcrystals are also wet by oil, as well as water. Although they are far more hydrophilic than they are lyophilic, it is theorized that the crystals do locate in the interface around the fat globules. For this reason, even though they do not have surfactant properties, they do possess excellent emulsion stabilization properties.

Besides the obvious emulsion, foam stabilization, suspension, and thickening functions of microcrystalline cellulose, the use of these insoluble crystallites is effective in the control of moisture migration and texture modification. The structuring effect of microcrystals are used extensively in whipped toppings, ice creams and formed foods such as extruded French fries.

B. Functions of Microcrystalline Cellulose in Food

Once the proper level of microcrystals to the available water is reached, the following functions can be realized:

**Adsorption of RC Crystallites
onto Oil Droplets**

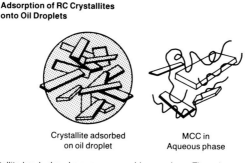

Crystallite adsorbed   MCC in
on oil droplet   Aqueous phase

Each RC crystallite has hydroxyl groups arranged in one plane. These have more an affinity for water
on one side and more an affinity for oil on the other. Thus, they line up on the o/w interface to physically
structure emulsion.

FIGURE 14.   MCC aqueous dispersion.

Function of Colloidal MCC (Cellulose Gel) in Foods

Opacifier	Provide cling
Suspended aid	Control moisture migration
Ice crystal control	Provide creaminess
Emulsion stabilizer	Flow control
Thixotropic thickener	Texture modification
Foam stabilizer	Dietary fiber
Heat stability	Structure of formed foods
Nonnutritive gelling agent	Alternate ingredient

1. Heat Stability

Two properties of colloidal microcrystalline cellulose are especially valuable in the preparation of heat-stable food products. These are emulsion stabilizer and the maintenance of product consistency during the application of high temperature processing. Colloidal microcrystalline cellulose (cellulose gel) functions as an emulsion stabilizer and thickener at high temperatures, because it forms a colloidal dispersion of solid particles when sheared in water. The physical properties of colloidal dispersions are quite different from the properties of starch gels, gum solutions, and other water soluble materials commonly used in food technology. In fact, products stabilized by colloidal particles have been used in industrial applications, but are now just being developed by the food industry.

Colloidal dispersions of microcrystalline cellulose have been found advantageous for food preparations where heat will be encountered in manufacturing and are used especially when food acids are present. Starch- or gum-based systems are often ruined under these conditions. In addition, microcrystalline cellulose is not discolored at high temperature or low pH.

Temperature changes have little effect on the viscosity of the microcrystalline cellulose dispersion (Figure 15). Microcrystalline cellulose colloidal dispersions can be either pourable suspensions or smooth thixotropic gels, depending on the concentration of the solids. A wide range of consistencies are obtained in food products containing microcrystalline cellulose.

2. Starch Extension

Using a ratio of three or four parts of starch to one part of microcrystalline cellulose (cellulose gel) will generally reduce the amount of starch thickener required by 20 to 25%. For example, a 5% modified starch can be replaced by a 3.75 blend of starch/microcrystalline cellulose. This blend should have most of the characteristics of the original starch system. Claims that have been made for microcrystalline cellulose in such applications are

●   Less masking of flavor

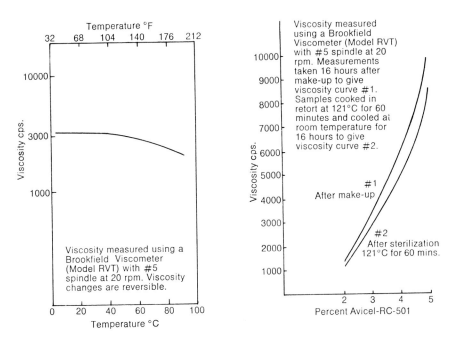

FIGURE 15.   Effect of varying temperatures on viscosity.

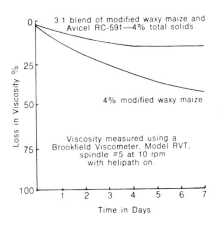

FIGURE 16.   Effect of prolonged heat on the viscosity of a waxy maize starch and a waxy maize/MCC.

- Improved flow control
- Greater resistance to breakdown and shearing
- Improved heat stability
- Improved stability in low pH foods

The graph in Figure 16 shows that under prolonged heat, the heat stability of a blend of microcrystalline cellulose and starch is better than that of a 4% modified waxy maize by itself. Admittedly, 77°C for 7 days is very severe — but so are ovens, steam tables, and vending machines.

Standardizing the viscosity, body, and texture of fillings such as those used in the baking industry is an excellent use of cellulose gel. Usually these sauces are overheated and then sheared during pumping transfers and reheated to baking temperatures, resulting in a variation

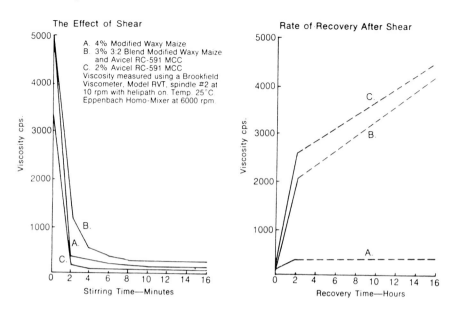

FIGURE 17.   Effect of shear on viscosity of starch systems.

of viscosity. Cellulose gel (microcrystalline cellulose) can standardize this loss as shown in Figure 17.

3. Foam Stability

Cellulose gel (MCC) is an excellent foam stabilizer, but it does not have film forming properties. When used in a topping, ingredients with film forming properties such as sodium caseinate or hydroxypropyl cellulose are required. A fat reduction of 3 to 4% is usually possible when cellulose gel is used.

Insoluble MCC microcrystals are used effectively in frozen desserts to stabilize the foam and improve overrun control (Figure 18). These microcrystals, located primarily in the aqueous phase, improve the extrusion characteristics and at the same time will strengthen the protein film around each air cell (Figure 19).

4. Sugar Gels

The addition of sucrose to aqueous dispersions of colloidal MCC yields higher viscosities than would be expected from the microcrystalline cellulose/water ratios (Figure 20). At concentrations of less than 1% of microcrystalline cellulose type RC-591, viscous pourable syrups are obtained; but at cellulose gel concentrations higher than 1%, thixotropic gels are formed. Thus it is possible to impart many unique rheological properties of microcrystalline cellulose to gelled sucrose systems (Figure 21).

5. Ice Crystal Control

The addition of 0.4% microcrystalline cellulose to ice cream mix preserves the original texture of the frozen dessert products during storage and distribution (Figure 22) by increasing their resistance to heat shock and by maintaining these products' three-phase system of air/fat/water. It also allows for reduction of fat and solids content by 2 to 4% with minimal loss of texture. It is theorized that as the water freezes, it forms large ice crystals which force concentration of the proteins and other solids, causing them to aggregate. But when the ice crystals rethaw, the solids do not fully rehydrate (Figures 23 and 24). Large surface areas of the microcrystalline cellulose allows for absorption of the moisture during the freeze/thaw cycle, thus preventing moisture migration and controlling aggregation of protein.

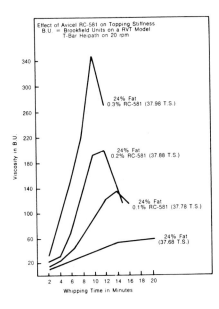

FIGURE 18. Effect of MCC on the stiffness of a vegetable fat topping.

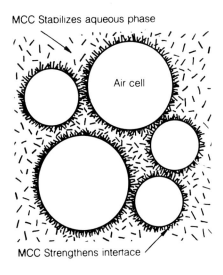

FIGURE 19. MCC in foams.

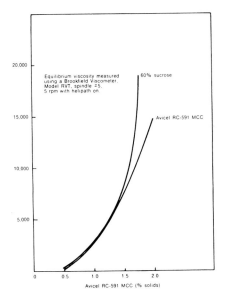

FIGURE 20. MCC/sugar syrup.

Like carrageenan, microcrystalline cellulose (cellulose gel) has the ability to prevent whey separation in mixes, thereby countering the destabilizing effects of some soluble gums (Figures 25 and 26). Illustrative of its multifunctional characteristics, the stabilizing effect of microcrystalline cellulose on serum solids is also related to its ability to control ice crystal growth during storage.

6. Thickening with Flavorable Mouthfeel

Colloidal MCC dispersions are capable of creating a mouthfeel which is unique. For example, it can be used to shorten the texture of a gum or starch combination or to control

FIGURE 21.   Photograph of bakery filling.

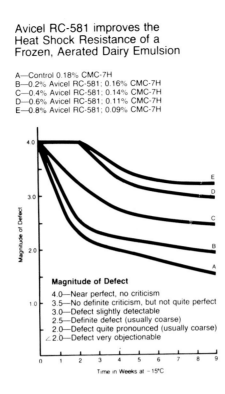

Avicel RC-581 improves the
Heat Shock Resistance of a
Frozen, Aerated Dairy Emulsion

A—Control 0.18% CMC-7H
B—0.2% Avicel RC-581; 0.16% CMC-7H
C—0.4% Avicel RC-581; 0.14% CMC-7H
D—0.6% Avicel RC-581; 0.11% CMC-7H
E—0.8% Avicel RC-581; 0.09% CMC-7H

Magnitude of Defect

4.0—Near perfect, no criticism
3.5—No definite criticism, but not quite perfect
3.0—Defect slightly detectable
2.5—Definite defect (usually coarse)
2.0—Defect quite pronounced (usually coarse)
<2.0—Defect very objectionable

FIGURE 22.   MCC improves heat shock resistance.

the flow of the food system without creating or affecting gummy or pasty texture. In many cases, these insoluble microcrystals are used to shorten the texture of xanthan gum solutions or frozen desserts containing large amounts of low DE corn syrup or whey. This is also attributed to the pseudoplastic characteristics of the insoluble microcrystals cellulose gel. Unlike many other vegetable gums, these insoluble microcrystals provide a clean mouthfeel and mask flavor only slightly. Microcrystalline cellulose gels would definitely fall into the "one" rating with starch and polysaccharide gums in Table 6.

7. Improved Clingability (Flow Control)

Figure 27 graphically demonstrates how the retention (clingability) of a starch sauce can be improved even at a temperature of 90°C. Note that the viscosity measurements on the samples containing 3% modified waxy maize and 0.5 to 1.5 cellulose gel (MCC) were lower than those of samples containing 5% of the same starch; yet, those containing the cellulose gel (MCC) had better retention.

8. Caloric Reduction and Fiber Addition

Cellulose gel (microcrystalline cellulose) is ideal for use in foods, where the caloric density must be controlled. In some foods, microcrystalline cellulose has also been used to increase the crude fiber content. The two basic types of MCC are both used for this application. The spray-dried agglomerates of MCC range in particle size from 90 to 200 μm. These powdered MCC products are used as a direct replacement for high caloric ingredients, such as flour, sugar, and fats in low-moisture food products.

In high-moisture food products, the multifunctional water dispersible colloidal MCC grades are applied. Not only do these water dispersible colloidal gels control the aqueous portion of the formulation, but they also suspend solids, stabilized foams and emulsions, modify texture, and control ice crystal growth in frozen desserts.

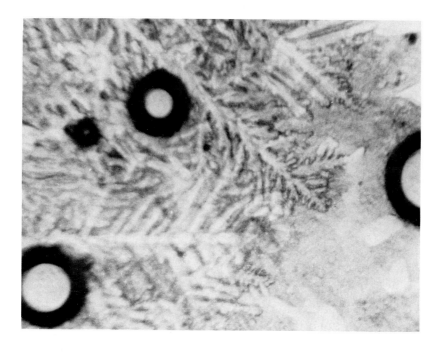

FIGURE 24.   Ice milk stabilized without MCC.

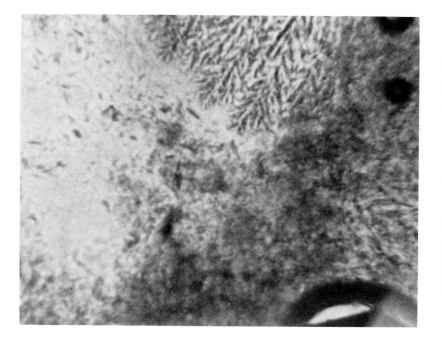

FIGURE 23.   Commercial sample ice milk with MCC.

FIGURE 25.   Ice milk stabilized with Avicel® RC demonstrates the ability of microcrystalline cellulose to stabilize solids.

Because of the removal of primary ingredients like sugar, fat, and starch, these applications are difficult and require proper selection of the cellulose types as well as alterations in the manufacturing procedure and the formulation.

9. Whitening
Even in low pH foods, microcrystalline cellulose will not brown on heating. The insoluble particles which are white are often used to opacify food systems.

10. Suspension
Dispersions of microcrystalline cellulose at levels of 0.3% will suspend solids such as chocolate in sterilized chocolate drinks, due to the interaction of the microcrystalline cellulose with the milk solids. Where this interaction is not available, levels of 0.5 to 1.5% may be required to suspend large heavy particles. An example might be in the canning operations, where suspensions of solids during cooking and retorting is important. For example, cooked starch granules can be suspended.

11. Emulsion Stabilization
Avicel® serves as an emulsion stabilizer in a way similar to other finely divided solids. Advantage may be taken of its physical properties to impart yield point and prevent coalescence of emulsified materials. Stokes' Law is overcome in emulsions which have a sufficient yield point. It is also theorized that because cellulose can be wet by both oil and water, that some of the crystallites locate in the O/W interface (Figure 14).

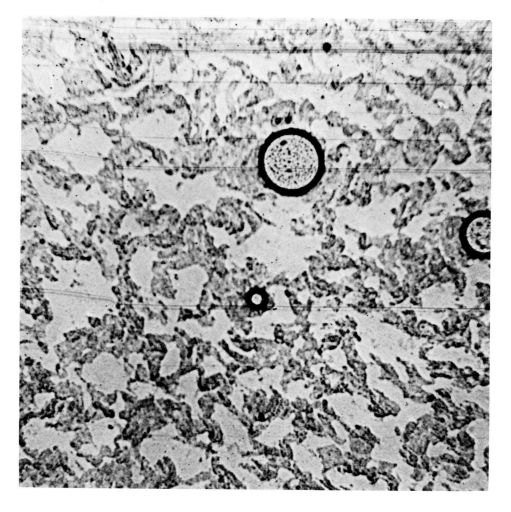

FIGURE 26.   Commercial sample of ice milk showing the flocculated milk solids.

12. Formed Extruded Foods

Probably the best example of MCC in an extruded food which utilizes the structuring effect of MCC is the fabricated frozen French fries. Major improvements in the quality of fabricated frozen French fries are achieved at all stages, from initial dough extrusion to final consumption, through the addition of 1% or less by weight of microcrystalline cellulose to the ingredient formulation (Figure 28).

Because it adds structural firmness and integrity to the dough, MCC (cellulose gel) improves extrudability and reduced breakage after extruding. This structural effect also improves the body and texture of the finished fry, providing a smoother consistency, fewer void spaces, and thinner crust. The result is a more tender, but firm fry with a more pleasing "mouthfeel".

A principal benefit achieved through the use of MCC (cellulose gel) is a reduction of approximately 15% in the absorption of cooking oil by the product during frying. MCC also maintains the fry structure retarding of moisture loss from the finished fry during storage under heat lamps, and keeping them firm and fresh-tasting over longer periods of storage after frying.

As MCC is added and the level increased, there is a corresponding increase in the firmness as indicated by these Universal Penetrometer readings (Figure 29). The normal life of French fries under heat lamp storage is approximately 4.5 min. Control French fries became limp

Table 6
CORRELATION BETWEEN MOUTHFEEL AND VISCOSITY VS. RATE-OF-SHEAR BEHAVIOR

Gum	Type	Concentration (%)	Organoleptic evaluation	
			Rating	How slimy
Starch	Cooked, corn	2.0	1	Non
Polysaccharide B-1459	—	0.15	1	Non
Sodium carboxypolymethylene	Sodium salt of Carbopol 934	0.3	2	Very slight
Carrageenan	SeaKem® 9	1.0	3	Somewhat
Gum karaya	—	1.0	3	Somewhat
Gum tragacanth	—	1.0	2	Very slight
Gum guar	—	0.6	2—3	Very slight to somewhat
Locust bean gum	—	0.7	3	Somewhat
Carboxymethylcellulose	CMC 7H XSP	1.0	4	Moderate
Sodium alginate	500 cps	1.3	5—6	Slimy to very
Low-methoxy pectin	Exchange® 466	5.0	7	Extremely
Methylcellulose	Methocel® 400	2.6	6	Very
Polyvinyl alcohol	Elvanol® 72—60	7.0	6	Very
Pectin	Exchange® 681	2.5	7	Extremely

From Szczesniak, A. S. and Farkas, E. H., *J. Food Sci.*, 27, 381—385, 1962. With permission.

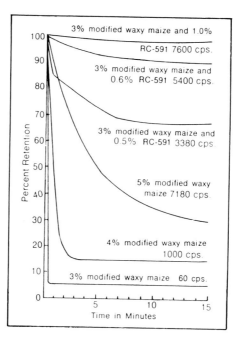

FIGURE 27.   Effect of MCC on clingability.

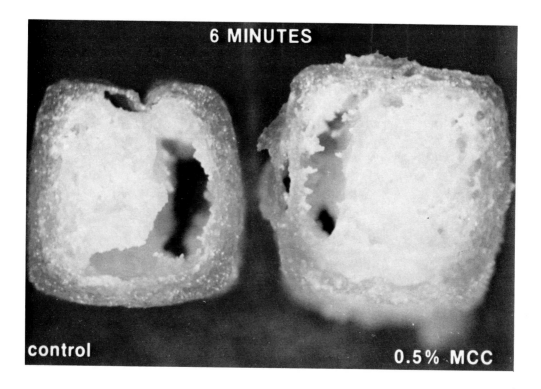

FIGURE 28.   Structuring effect of 0.5% MCC in frozen fabricated French fried potatoes.

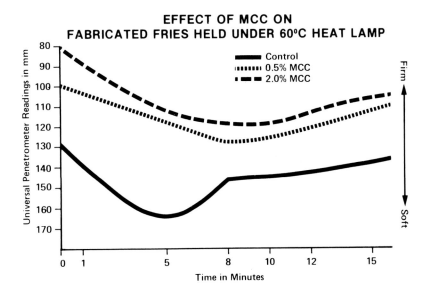

FIGURE 29. Firmness of extruded French fry.

within the 5-min period reaching their lowest point at approximately 4.5 min. However, fries containing MCC do not reach their lower point until 8 to 10 min and never reach the depth of limpness of the control. This firmness also helps to control breakage.

The functional properties of MCC which have found significant utility in food applications are listed in Table 7.

V. FEDERAL REGULATIONS PERTAINING TO THE USE OF MICROCRYSTALLINE CELLULOSE IN FOODS

FDA Status — Microcrystalline cellulose (cellulose gel) is generally recognized as safe (GRAS) by qualified experts and in accordance with the Food and Drug Administration.

Food Chemicals Codex Listing — Microcrystalline cellulose has been listed in the 4th supplement of the Food Chemicals Codex, first edition, by the National Academy of Sciences-National Research Council — as; cellulose, microcrystalline (cellulose gel).

Labeling — To label colloidal microcrystalline cellulose (cellulose gel) properly, the microcrystalline cellulose and the sodium carboxymethylcellulose must be declared. Synonyms listed in the Food Chemicals Codex for microcrystalline cellulose and sodium carboxymethylcellulose are cellulose gel and cellulose gum, respectively.

Table 7
FOOD APPLICATION OF MICROCRYSTALLINE CELLULOSE
(MCC)

	Type	Level (%)	Function
Grain milled and bakery products			
High fiber bread	PH-101	5.5	3/7
Low-cal cake	PH-101	4.8	3
Cheesecake	RC-591	1.7	20/8/29
Icings	RC-591	0.3	19/27/28/30
Bakery fillings (pectin)	RC-591	1.0	18/19/29/8
Bakery fillings (starch)	RC-591	1.25	18/26
Low-calorie cheesecake	RC-591	1.7	3/22
Fruit bar filling	RC-591	1.0	9/18/22/27
Chocolate cake sauces	RC-581	0.9	24/22/27
Lemon pie filling	WC-595	4.5	20/8/19
Meringue	RC-591	0.3	17
Chocolate pie filling	RC-591	0.75	20/8/19
Reduced calorie bread	PH-101	5.5	3
Low-cal cookies	PH-101	9.3	3
Cake	WC-595	2.0	8/17/28/23
Pancakes	WC-595	2.0	8/17/23/27
Canned specialties			
Meat salads	RC-501	3.0	18/14/22
Dips and spreads	RC-591	2.00—4.5	24/14/18/27
Aseptic puddings	RC-591	1.0	26/22/8/18
Potato salad	RC-591	3.5	14/22/18
White sauces	RC-591	1.0	10/16/14
Macaroni and cheese	RC-581	0.75	8/19
Creamed chipped beef	RC-591	1.0	10/16/14/30
Chicken à la king	RC-591	1.0	10/16/14/30
Heat stable sour cream	RC-501	5.5	18/14/8/20
Cream soups	RC-591	0.75	10/14/8/30
Clam chowder	RC-591	0.75	10/14/30
Dairy and imitation dairy products			
Imitation mozzarella cheese	PH-101	3.0	8/5/23/4
Frozen whipped toppings	RC-501/581	0.2—0.5	30/17/8
Whipped topping bases	RC-501/581	0.2—0.5	30/17/18
Aerosol topping	RC-501	0.3—0.5	30/17/18
Sterilized whipping cream	RC-591	0.3	14/17
Ice cream	RC-581	0.2—0.4	13/17/8/30
Ice milk	RC-581	0.3—0.6	13/17/8/30
Frozen fudge bars	RC-581	0.25	8/13/30
Imitation ice cream	RC-581	0.2—0.3	8/13/30/17
Imitation ice milk	RC-581	0.3—0.6	8/13/30/17
Milk shakes	RC-581	0.3—0.75	16/17
Soft serve ice milk	RC-581	0.2—0.5	19/27
Low-calorie frozen dessert	RC-501/581	0.35—1.25	13/8/3/19
Imitation milk	RC-581	0.4	10/19/30/12
Sterilized drinks (retorted aseptic)	RC-591	0.3	12/10/19/30
Imitation sour cream	RC-591	0.2—0.4	8/19
Frozen yogurt	RC-581/591	0.35	12/19/8/30
Dry ice mixes	WC-595	3.0	17/30/8/19
Frozen mousses	RC-591	0.8—1.5	8/30/17
Sherbet	RC-581	0.35	19/30
Beverages			
Sterilized dairy drinks	RC-581	0.3	14/19/12/30
Fruit beverages	RC-591	0.6	12/16/10/9
Canned emulsion beverages	RC-501	0.7	14/19/16/30

Table 7 (continued)
FOOD APPLICATION OF MICROCRYSTALLINE CELLULOSE (MCC)

	Type	Level (%)	Function
Beverages (continued)			
Protein-fortified beverages	RC-591	0.5	19/12/16/30
Fortified dietetic beverages	RC-591	0.5	20/16/12/30
Eggnog with alcohol	RC-591	0.75	19/10/30
Bloody Mary without alcohol	RC-591	0.25	12/19
Piña colada	RC-591	0.5	10/24/14/30
Sauces and salad dressing			
Tomato sauces	RC-581	0.25	16/24/18/9
Pourable salad dressings	RC-591	1.5	27/24/20/14
Low-calorie pourable	CL-611	2.5	27/14/20/21
Imitation mayonnaise	RC-591	2.5	20/30/19
Imitation salad dressings	RC-591	2.5	20/30/19/23
White sauce	RC-591	1.0	10/14/24
Barbecue sauce	RC-591	0.33	24/9/27/26
Cole slaw dressing	RC-591	2.5	27/24/20/21
Bernaise sauces	RC-591	1.5	10/14/24
Dry blended sauces	WC-595	4.0	10/14/24
Confection and jelly products			
Low-calorie candy	RC-591/PH-105	1.5—6.0	3/21
Imitation jelly	RC-591	0.75	8/9/29
Low-calorie jelly	RC-591	0.1	21/8/9/19
Marshmallow topping	RC-581	0.75	10/17/27
Low-calorie ice cream topping	RC-591		
Dry applications			
Grated and shredded cheese	PH-101	1.5	5
Enzyme tablet	PH-101	1.5	1
Salt tablet	PH-101	3.5	1
Proportioned nutrient tablet	PH-101	50/50	1
Flavor	RC-591/PH-101	50/50	2/15
Oleoresins	RC-591/PH-101	50/50	2/15
Food acids	PH-101	50/50	2
Color	RC-591/PH-101	50/50	2/15
Shortening	RC-591/PH-101	50/50	2/15
Dry mixes	PH-101	1.0	5
Peanut butter	PH-101	50/50	2
Dry whipped toppings	WC-595	2.0	16/17/30
Dry puddings	WC-595	2.0	8/16/30
Formed foods			
Frozen French fries	RC-591	0.4	28/8/4
Frozen sweet potato patty	RC-591	2.75	4/8/26
Frozen fish sticks	RC-591	0.5	4/8/26
Crab patty with soya	RC-591	0.5	8/4/23
Expanded snacks	PH-101	1.0	25/4
Simulated fruit pieces	PH-101	1.1	9/8
Soya analogs	RC-591/PH-101		4/18/14
Imitation spices	PH-101	50/50	9/25/2
Miscellaneous			
Low-calorie solid foods	PH-101	1.0—11.0	3
Low-calorie food gels	RC-591	0.75—2.0	20/21
Dehydrated foods	PH-101	5.0	11/25
Batter coatings	WC-595	2.5	28/25/24

Table 7 (continued)
FOOD APPLICATION OF MICROCRYSTALLINE CELLULOSE
(MCC)

Key: MCC Functions

1. Tableting binder and disintegrant	16. Thickener
2. Carrier	17. Foam stabilizer
3. Nonnutritive bulking	18. Heat stabilizer
4. Binder	19. Stabilizer
5. Anticaking agent	20. Thixotropic agent
6. Flow aid	21. Nonnutritive gel agent
7. Dietary fiber	22. Gelling agent
8. Texture modification	23. Alternative ingredients
9. Cellulose component	24. Cling
10. Opacifier	25. Quick-drying
11. Hydration aid	26. Flavor release
12. Suspending agent	27. Flow control
13. Ice crystal control	28. Moisture migration control
14. Emulsion stabilizer	29. Cost reduction
15. Microdispersant	30. Creaminess

MCC Type	FMC Corporation Designation
Powder cellulose	PH-101
Bulk-dried	RC-501/RC-581
Spray-dried	RC-581/RC-591
75% whey/25% MCC	WC-595

REFERENCES

1. **Anon.,** Crystalline cellulose offers many advantages as a food ingredient, *Food Process. (Chicago),* November 1960.
2. **Trauberman, L.,** Crystalline cellulose: versatile new food ingredient, *Food Eng.,* August 1961.
3. **Wright, R. H.,** New diet breakthrough, *Family Digest,* p. 41, November 1961.
4. **Young, W. R.,** Food that isn't food, *Life,* 50, 22, 1961.
5. **Young, W. R.,** New food that can't fatten you, *Readers Digest,* p. 64, September 1961.
6. **Battista, D. A.,** Food Compositions Incorporating Cellulose Crystallite Aggregates, U.S. Patent 3,023,104, 1962.
7. **Anon.,** Diversify the dietetics, *Food Eng.,* May 1962.
8. **Herald, C. T.,** Microcrystalline cellulose — new ingredient could cut candy calories if non-nutritive problems are overcome, *Manuf. Confect.,* May 1962.
9. **Winchester, J. H.,** If you love to eat and hate to diet . . . *Readers Digest,* January 1963.
10. **Fassnacht, J. H. and Bower, F. A.,** Food formulations adopted to aerosol packaging, *Food Technol.* 19(8), 44—48, 1965.
11. **Wyden, P.,** Coming soon: the new non-foods, in *The Overweight Society,* Pocket Cardinal Edition, 1966, 234—238.
12. **Kirk, J. K.,** Proposal to amend identity standards by listing MCC as an optional ingredient in ice cream and fruit sherbet, *Fed. Regist.,* 31, 158, 1966.
13. **Herald, C. T.,** Non-nutritives herald new era for industry, *Candy Industry,* September 1966.
14. **Herald, C. T., Raynor, G. E., Jr., and Klis, J. B.,** New colloidal cellulose . . . for heat sterilized salads, *Food Process. Mark.,* p. 54, November 1966.
15. **Anon.,** Stabilizer . . . makes possible canned salads, other new food concepts, Putnam Food Award, *Food Process. Mark.,* p. 54, November 1966.
16. **Anon.,** Tuna fish salad in a can? No longer impossible, *Chemmunique,* 16(1),2, 1967.
17. **Galliker, L. G.,** Ice milk sales increase ten-fold, *Ice Cream Rev.,* December 1967.

18. **Anon.,** Frozen desserts: definitions and standards of identity, *CFR Title 21 — Food and Drugs*, Subchapter B. Part 20 — Frozen Desserts, corrected through 1968.
19. **Knightly, W. H.,** The role of ingredients in the formulations of whipped toppings, *Food Technol.*, 22, 731, 73, 1968.
20. **Pearson, A. M.,** Microcrystalline cellulose in ice cream and ice milk, O.A.I.C.M. Annual Convention, April 2, 1968.
21. **Pearson, A. M.,** What can microcrystalline cellulose do for ice cream, sherbet, ice milk?, *Can. Dairy Ice Cream J.*, p. 21, August 1968.
22. **Lee, C. J., Rust, E. M., and Reber, E. F.,** Acceptability of foods containing a bulking agent, *J. Am. Diet. Assoc.*, 54, 210—214, 1969.
23. **Shurpalekar, K. S., Sandaravalli, O. E., and Narayanara, M.,** *Effect of Inclusion of Extraneous Cellulose of the Gastric Emptying Time in Adult Albino Rats*, Physiological Study from India, September 1968.
24. **Glicksman, M. H.,** Microcrystalline cellulose, in *Gum Technology in the Food Industry*, Academic Press, New York, 1969, 403—412.
25. **Anon.,** Cellulose plus starch improves tomato sauces . . . , *Food Process. (Chicago)*, November 1970.
26. **McCormick, R. D.,** Control of viscosity and emulsion stability in foods using modified microcrystalline cellulose, *Food Prod. Dev.*, 4, 4, 19—26, 1970.
27. **Anon.,** Cellulose microcrystalline (cellulose gel), description, identification, etc., *Fourth Supplement to the Food Chemicals Codex First Edition*, February 1971, 2—4.
28. **Walker, G. C., Tuomy, J. M., and Walts, C. C.,** Investigation of Factors Affecting the Melt Down of Soft Serve Imitation Ice Milk, Tech. Rep. 71-29-FL, March 1971.
29. **Pratt, D. E., Reben, E. F., and Klockow, J. H.,** Bulking agents in foods, *J. Am. Diet. Assoc.*, 59(2), 120, 1971.
30. **Keeney, P. and Josephson, D. V.,** Better heat shock resistance and extrudability in ice creams with microcrystalline cellulose, *Food Prod. Dev.*, November 1972.
31. **Nollman, D. S. and Pratt, D. F.,** Protein concentrates and cellulose as additives in meat loaves, *J. Am. Diet. Assoc.*, 61, 658—661, 1972.
32. **Peleg, M. Mannheim, C. H., and Passey, N.,** Flow properties of some food powders, *J. Food Sci.*, 38, 959—964, 1973.
33. **Fine, S. D.,** Grated cheeses: identity standard; microcrystalline cellulose as optional anti-caking agent; confirmation of effective date, *Fed. Regist.*, 38, 49, 111, 1973.
34. **Sandaravalli, O. E., Shurpalekar, K. S. and Narayanara, M.,** Inclusion of cellulose in calorie-restricted diets, *J. Am. Diet. Assoc.*, 62, 41—43, January 1973.
35. American National Standard K6538 1973 (R-1967), Standard Method of Test for Intrinsic Viscosity of Cellulose, by *American National Standards Institute*, January 18, 1973.
36. **Mussellwhite, P. R.,** Avicel and Emulsion Stability, presentation at Distributors Meeting, May 1974.
37. Cellulose gel flowable for cheeses, *Food Process. (Chicago)*, May 1974.
38. **Herald, C. T. and Raynor, G. E., Jr.,** *Colloidal Microcrystalline Cellulose in Food*, Memo, unknown date.
39. Unknown, Bulk of evidence favors fibers, *Food Process. (Chicago)*, p. 70, October 1974.
40. **McCormick, R. D.,** Controlling heat stability for bakery fillings, toppings, *Food Prod. Dev.*, December 1974.
41. **Levine, M. B. and Potter, N. N.,** Freeze-thaw stability of tomato slices; effects of additives, freezing and thawing rates, *Food Prod. Dev.*, November 1974.
42. Food products/operations improved through innovative application of ingredients, *Food Process. (Chicago)*, p. 80—81, June 1975.
43. **Vos, P. T. and Labuza, T. P.,** Technique for measurement of water activity in high Aw range, *Agric. Food Chem.*, 22(2), March/April 1974.
44. **McGinley, E. J. and Thomas, W. R.,** *Microcrystalline Cellulose; An Anti-Caking Agent for Grated and Shredded Cheese*, paper given at 12th Annual Marschall Invitational Italian Cheese Seminar, April 28 to 29, 1975.
45. **Andres, C.,** Stabilizers 2 — Gums, *Food Process. (Chicago)*, January 1976.
46. **Lah, N., Karel, M., and Flink, J. M.,** A simulated fruit gel suitable for freeze dehydration, *J. Food Sci.*, Vol. 41, 1976.
47. **Stockman, S. A.,** FMC's Avicel® solves sticky problem of caking in high moisture cheeses, *Food Eng.*, January 1976.
48. **McCormick, R. D.,** Micro dispersion — versatile system for selective release of functional ingredients, *Food Prod. Dev.*, July to August 1976.
49. **Brys, K. and Zabik, M.,** Microcrystalline cellulose replacement in cakes and biscuits, *J. Am. Diet. Assoc.*, 64, 1, 1976.
50. **Pomerantz, Y. et al.,** Fiber in Breadmaking — Effects on Functional Properties

51. **Warmbier, H. C. et al.,** Non-enzymatic browning kinetics in an intermediate moisture model system: effects of glucose to lysine ratio, *J. Food Sci.,* 41, 1976.

52. **Andres, C.,** MCC Overcomes Problems in Dietetic and Low-Sugar Gel Systems.

53. Microcrystalline cellulose secret to ovenproof fillings, *Canner/Packer,* November 1976.

54. **Gejl-Hannsen, F. et al.,** Application of microscopic techniques to the description of structure of dehydrated food systems, *J. Food Sci.,* 41, 1976.

55. **Collins, J. L. and Falasinnu, G. U.,** Dietary fiber as an ingredient in cookies, *Tenn. Farm Home Sci. Prog. Rep.,* 101, January, February, March 1977.

56. **Arbuckle, W. S.,** Advances in Ingredients and Technology in Ice Cream Manufacture, source unknown.

57. **Satin, M. et al.,** Design of a Commercial Natural Fiber White Bread, paper given at AACC Meeting in San Francisco, California, October 25, 1977; Fiber Bibliography, No. 468 (Section 4).

58. **Praha, P. O.,** Use of microcrystalline cellulose in the production of pastries, *Prom. Potravin,* 21(10), 306, 1970.

59. **Andres, C.,** Highly porous, inert carrier aids liquid ingredients handling, *Food Process. (Chicago),* May 1978.

60. **Steege, H., Phillipp, B., Engst, R., Magister, G., Lewerence, H. J.,and Bleyl, D.,** Microcrystalline cellulose powders: properties and possible applications in nutrition, *Tappi,* 61(5), 1978.

Chapter 2

SODIUM CARBOXYMETHYLCELLULOSE (CMC)

John D. Keller

TABLE OF CONTENTS

I. HISTORICAL BACKGROUND

Sodium carboxymethylcellulose or CMC was developed in Germany during the World War I era as a potential substitute for gelatin.[1,2] However, technical difficulties and high production costs stifled and delayed full-scale commercialization of CMC at that time.

In 1935, it was discovered that sodium carboxymethylcellulose improved detergency.[3] Specifically, CMC was used in synthetic detergents to prevent the undesirable phenomenon of "tattletale gray" caused by soil redeposition on fabric during normal washing and rinsing. As a point of interest, an accepted mechanism of this prevention is that CMC is absorbed on the fabric surface (e.g., cotton or rayon) by hydrogen bonding. The anionic polymer imparts an electronegative charge to the cloth which repels the dirt that is also negative.[4] Another hypothesis suggests that the colloidal system developed by the CMC polymer deters the reestablishment of particulate soil on fabric surfaces.

Prior to World War II, fatty acid soaps were used extensively to clean clothes. The discovery that CMC improved the efficiency of synthetic detergents (which rapidly replaced natural cleaners due to channeling of fatty acid supplies to military use) coupled with a shortage of water-soluble gums in wartime Germany (W.W. II) stimulated renewed interest in the manufacture of this material. The first commercial CMC was made by Kalle and Co. at Wiesbaden-Biebrich during this period, i.e., the late 1930s.

Concurrent interest in CMC evolved in the U.S. when Hercules, Inc. developed a commercial process in the fall of 1943.[5] Full-scale commercial production was realized by Hercules at Hopewell, Va. in 1946.

With the close of the world conflict in 1945, CMC began finding many new and varied applications in products requiring water control. The first food use for CMC (ice cream) arose from the gelatin shortage just after World War II. CMC was tried and found to be an effective substitute. By 1950 to 1952, the use of CMC in ice cream was well established. Since then, the use of CMC in food products, as well as pet food, has diversified and grown substantially. Outside of foods, the use of CMC in paints, pharmaceuticals, cosmetics, toothpaste, petroleum, paper, cements, adhesives, ceramics, and textiles has advanced markedly. Table 1 describes a general use pattern for CMC in the world marketplace.[6]

Today, the ice cream stabilizer area still represents one of the largest food uses for CMC.

II. GENERAL DESCRIPTION

Sodium carboxymethylcellulose is a linear, long-chain, water-soluble, anionic, man-modified polysaccharide. It is a polymer which enjoys premier membership in the category of substances best described as "chemically modified natural gums". Appellations such as cellulose ether or cellulose derivative, as well as a myriad of trade names, all refer to the material known universally as CMC.

Highly purified CMC required for use in the food, pharmaceutical, and cosmetic industries is designated as cellulose gum. The accepted definition of cellulose gum (see Section VI, Regulatory Status) has been set forth by the U.S. Food and Drug Administration (FDA). This governing agency has listed purified sodium carboxymethylcellulose (cellulose gum) among those materials recognized as GRAS, i.e., "generally recognized as safe." Additionally, the Food Chemical Codex (FCC) and the Food and Agriculture Organization (FAO) of the United Nations have established specifications for the identity and purity of sodium carboxymethylcellulose for food use worldwide. Table 2 provides a classification of various CMC grades according to purity, including the purified premium grade known as cellulose gum.

Purified sodium carboxymethylcellulose is a white- to cream-colored, tasteless, odorless, free-flowing powder. This latter characteristic is maintained only by adequate packaging, as the material readily absorbs atmospheric moisture.

Table 1
WORLDWIDE USE PATTERN FOR
CMC

Industrial category	%
Detergents	38
Petroleum	16
Paper	10
Food	7
Textiles	5
Pharmaceuticals, cosmetics, toothpaste	4
Building materials	4
Mining	2
Others	14

Table 2
PURITY CLASSIFICATION OF CMC GRADES

Class	Purity range % Na CMC	General use
Purified[a]	99.5 +	Food, pharmaceutical, cosmetic, toothpaste
Technical or refined	96—99	Industrial uses
Semitechnical	70—90	Detergent, mining, petroleum
Low assay or crude	50—70	Detergent, mining, petroleum

[a]   Only grade referred to as cellulose gum.

CMC is perhaps one of the most versatile and widely used water-soluble polymers in existence. The material is a useful rheological tool that has undergone rapid growth and expansion into many facets of industry, particularly foods. The commercial value of CMC in the U.S. is readily demonstrated by a steady increase in production of the gum from the late 1940s, when it was first introduced, to present date (Figure 1). It is anticipated that the food, pharmaceuticals, cosmetics, and toiletries markets for CMC will continue a steady 4 to 5% growth per annum based on recent industrial trends.[8]

Ironically, the largest single market for CMC, detergents, is believed to have peaked and is expected to show a slight decline in years to come.[8] Utilization of CMC in the petroleum field, particularly drilling muds, peaked in 1981 when the active drilling rig count sky-rocketed. Continued growth of CMC usage in the petroleum field will obviously depend on future world energy needs, which are uncertain.

Drilling muds are fluids which cool and lubricate the drill bit, transport cuttings to the surface, and function as a sealant on the walls of the borehole. CMC thickens the mud and thus decreases loss of the fluid into surrounding rock strata.

In foods, the basic properties of CMC that enhance its commercial value are its ability to thicken water, act as a moisture binder, dissolve rapidly in both hot and cold aqueous systems, and "texturize" a wide range of food products. It is tasteless, odorless, and uniquely forms crystal clear solutions, whereas other gums often impart cloudiness or opacity due to the presence of significant impurities or insolubles such as protein, lipids, etc. Cellulose gum is physiologically inert and noncaloric, because it is not metabolized by the human digestive system. This latter property makes it particularly useful in dietetic foods. In other food systems, sodium carboxymethylcellulose serves as an extrusion aid, acts as a binder, helps stabilize emulsions, and retards sugar crystal growth. In frozen desserts, it prevents ice crystal growth and phase separation. The gum is compatible with a wide range of other

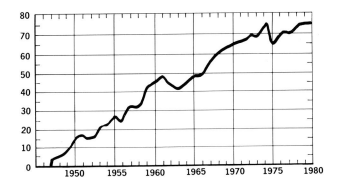

FIGURE 1.   Carboxymethylcellulose production in the U. S. (1947—1980).

food ingredients, particularly proteins, sugars, and other hydrocolloids where synergistic interaction can occur.

Sodium carboxymethylcellulose is available in several different viscosity grades, particle size grades, special rheological grades, and combinations thereof which permit tailor-made application of the gum to a multitiude of food systems. This versatility was recognized by Whistler[9] prior to development of the many different and sophisticated CMC types that now exist — "It is conceivable that as more is learned about the relationship of structure to the physical properties of polymers, specific gum properties will probably be custom-tailored into starch and cellulose molecules so that the properties of the custom-made products will more closely match the properties desired in special gum applications."[9]

As mentioned previously, sodium carboxymethylcellulose is a "chemically modified natural gum". Man's ability to control the modification of cellulose into a versatile gum product and a nearly inexhaustible raw material supply provides CMC with a distinct advantage over totally natural gums which are limited in the above respects.

Traditionally, naturally occurring gums derived from seaweed, trees, seeds, and similar sources were employed in foods. In recent years, this tradition has progressively diminished. Substitutes are now replacing the all-natural gums which have decreased in output and risen sharply in price because of geographical, climatic, economic, and political influences. CMC is largely unaffected by these drawbacks.

Glicksman[10] observed that chemically modified natural gums are still in their "infancy", but "are steadily pressing at the position of natural gums and enlarging their foothold in the field as newer and better modified hydrocolloids become available".[10] Today, this situation has become reality as cellulose gum (by volume) is one of the most widely used single gums in foods. One can cite (1) uniformity of properties and specifications, (2) unlimited availability of raw materials, (3) versatility, (4) purity, and (5) relatively stable prices (although energy costs have affected pricing more dramatically in recent years) as factors which have caused the evolution of "chemically modified natural gums" like CMC into the limelight of the food industry.

III. MANUFACTURE

A. Raw Material

Cellulose is nature's raw material. It is the main structural component of most land plants and is perhaps the most abundant natural resource on earth. Cellulose is the fundamental backbone of CMC.

The cellulose molecule is a straight chain polymer composed of anhydroglucose units (Figure 2). In this structure, "n" is the number of anhydroglucose units which are joined

STRUCTURE OF CELLULOSE

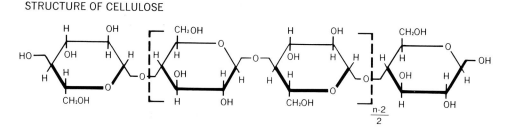

FIGURE 2.   Cellulose molecule.

through beta-1,4 glucosidic linkages, or the degree of polymerization (DP) of the cellulose. The DP may vary from 100 to 3500 units, which makes cellulose a high molecular weight material. Two linked anhydroglucose residues (in brackets) constitute a cellobiose unit.

Each anhydroglucose unit contains three hydroxyl groups. These are the reaction sites where the carboxymethyl group becomes attached by ether linkage.

Cellulose is a fibrous solid that is water insoluble. This characteristic is attributed to alignment of the long thread-like molecules into fibers where highly ordered crystalline regions exist due to association by hydrogen bonding. Manners[11] noted that the chemical and physical properties of cellulose are largely dependent on the number, size, and arrangement of the crystalline regions. It is the strong mechanical forces of the crystalline regions that keep the network of cellulose fibers intact and disallow the entrance of water between polymer chains to associate with hydroxyl groups, giving the desired property of solubility. Baird and Speicher[12] confirmed that modification of the cellulose through introduction of a controlled amount of sodium carboxymethyl groups ($Na^+-OOC-CH_2$) onto the molecule imparts water solubility. These constituents help to hold the chains apart, allowing water to enter and achieve dissolution. This will be discussed in greater depth when solution properties are covered.

Chemical cotton and wood pulp are the chief commercial sources of raw material cellulose used for the manufacture of CMC. The chemical cotton is derived from cotton linters (a by-product of the cotton industry).

Thanks to the inventive prowess of Eli Whitney, cotton seeds are removed from the cotton ball by the cotton gin. These seeds are then sent to the oil manufacturers for pressing out of the cottonseed oil. However, before doing so, the fine fuzz or lint on the seeds is removed by cutters to give the material utilized for CMC manufacture.

Cotton linters are the purest form of natural cellulose available, consisting of about 98% α-cellulose. Wood contains about 40 to 50%. Other materials, such as corn stalks, corn cobs, and wheat straw, contain about 30% cellulose and offer a potential raw material source for cellulose, but have not been utilized.[13]

B. Manufacturing Process

A variety of techniques (most of which are proprietary) have been utilized by different manufacturers to make CMC. All of these processes rely upon a conventional chemical reaction (etherification) which remains unchanged from that set forth by the inventor.[2] Figure 3 outlines the basic chemistry of manufacture which takes place in two steps. Purified cellulose is first steeped in strong sodium hydroxide to swell open the tightly bound cellulose chains mentioned earlier. The extent of swelling depends on the starting material, i.e., the degree of crystalline regions within the lattice network of the cellulose. Once swollen, solvent, containing sodium monochloroacetate (MCA), is allowed to enter and complete the substitution of the carboxymethyl group onto one of three reactive hydroxyl sites. The process is conducted under rigidly controlled conditions which may vary depending on the quality of the product desired. By-products of the reaction are sodium glycolate and sodium chloride which are removed by extraction to reproduce the more purified grades of CMC.

MANUFACTURE OF CMC

$$R - OH + NaOH \longrightarrow R - ONa + H_2O \text{ (1ST)}$$

$$R - ONa + CL - CH_2 - COONa \longrightarrow R - O - CH_2 - COONa + NaCL \text{ (2ND)}$$

FIGURE 3. Manufacture of cellulose gum.

Figure 4 shows an idealized unit structure for sodium carboxymethylcellulose with a DS of 1.0. The DS stands for the "degree of substitution" or the average number of hydroxyl groups substituted per anhydroglucose unit. If all three hydroxyls on the anhydroglucose monomer were substituted, the DS would be 3.0. Theroretically, this is the maximum DS one could attain, but in practice it is impossible to achieve. Most available commercial products have a degree of substitution ranging from 0.4 to 1.4. Food grade sodium carboxymethylcellulose is presently restricted to a maximum DS of 0.95, but no lower limit exists. A DS of 0.7 is the most commonly employed DS level in food systems.

The literature routinely depicts the 6-position hydroxyl as the primary site for substitution of the carboxymethyl group as seen in the Haworth structure of Figure 4. However, NMR studies by Reuben and Connor[14] have revealed that the reactivity of the hydroxyls on the anhydroglucose ring of cellulose towards carboxymethylation decreases in the order of 2>6>3. So in fact, the 2-position is the primary reaction site, followed by the 6-position, with the 3-position the least preferential substitution site. The reason — it is hypothesized that the 2-position has the greatest amount of electronic pull in the etherification reaction. Whether or not this hypothesis holds true for other cellulose derivatives is unknown, since they have not been studied as extensively as CMC.

Reuben and Connor are in agreement with earlier findings by Spurlin[15] where a statistical kinetic model assumed that substitution patterns on cellulose derivatives are governed by relative rate constants for reaction of the three hydroxyl groups on the glucose residue.

C. Types of CMC

General estimates indicate that over 30 producers of CMC exist worldwide. In aggregate, these manufacturers are responsible for the promulgation of over 300 different types of CMC into the industrial marketplace. Even so, less than one third of the world manufacturers produce material that meets the purity requirements for food grade CMC.

CMC gained entrance into the food industry (ice cream) when a basic grade of the material was effectively substituted for an established stabilizer (gelatin) that was in short supply. With the development of more sophisticated and complex food systems over the years, the need for more specialized CMC types increased rapidly to meet new functionality demands of a growing food industry.

Through manipulation of various combinations and permutations of the chemical and physical properties that characterize CMC — degree of substitution (DS); degree of polymerization (DP); uniformity of substitution; raw material source; particle size, shape, and density — manufacturers now produce a vast array of special types and grades of cellulose gum which fulfill numerous food application requirements. Table 3 lists the different DS-viscosity types of CMC available from one producer. Generally, the "type" is approximately 10 times the DS.

IV. CHEMICAL AND PHYSICAL PROPERTIES

A. Solution Properties

1. Solubility

CMC is soluble in either hot or cold water, which facilitates use of the hydrocolloid in

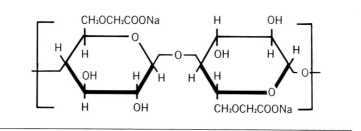

FIGURE 4.   Idealized unit structure of cellulose gum, with a DS of 1.0.

Table 3
SOME VISCOSITY CODES FOR HERCULES CELLULOSE GUM

Viscosity range in centipoises at 25°C[a]	Designations for indicated substitution types			
	4	7	9	12
High — at 1% concentration				
2500—5000		7H4F	9H4F	
1000—2800		7H3SF, 7HOF		
1500—3000		7HF		
500—1200	4H1F			
Medium — at 2% concentration				
800—3100			9M31F	
400—800		7MF	9M8F	
200—800		7M8SF		
300—800				
100—200		7M2F		
50—100		7M1F		
Low — at 2% concentration				
25—50		7LF		
20 max				

[a]   Ranges shown in this table are not necessarily current specifications.

a wide variety of food applications and processes. It is insoluble in pure organic solvents. However, solubility occurs in mixtures of water and water-miscible organic solvents such as short-chain alcohols (e.g., ethanol). Table 4 shows the tolerance of various molecular weight grades of CMC to a mixed solvent system. The low-viscosity types are considerably more tolerant to increasing ethanol concentration than higher-viscosity types. CMC solutions of low concentration can be made with up to 50% ethanol or 40% acetone. Since ethanol is used to precipitate most gums during manufacture, the excellent tolerance of CMC to considerable ethanol concentration in a mixed solvent is unique. This property becomes very important in the application of CMC as a stabilizer to alcoholic beverages and instant bar mixes where clarity and viscosity are needed.

Sodium carboxymethylcellulose, as the name implies, is a salt of a carboxylic acid. A dilute solution of CMC is characterized by neutral pH (about 7) and has virtually all of its carboxylic acid groups in the sodium salt form and very few in the free acid form. When solution pH is adjusted to 3.0 or lower, sodium carboxymethylcellulose reverts from the salt form to the free acid form (carboxymethylcellulose) which is insoluble. The pK of

Table 4
TOLERANCE OF HERCULES
CELLULOSE GUM SOLUTIONS FOR
ETHANOL

Cellulose gum type	Volume ratio of ethanol to cellulose gum solution (1%)	
	First evident haze	First distinct precipitate
7L	2.4 to 1	3.6 to 1
7M	2.1 to 1	2.7 to 1
7H	1.6 to 1	1.6 to 1

Note: In these tests, ethanol (95%) was added slowly, at room temperature, to the vigorously stirred 1% cellulose gum solution.

sodium carboxymethylcellulose ranges from 4.2 to 4.4 and varies somewhat with the degree of substitution.

CMC, like many other water-soluble substances, may be salted out of solution with alkali metal salts. However, because of the very hydrophilic nature of the polymer, CMC is much more tolerant of many types and concentrations of salt than other hydrocolloids such as carrageenan, alginates, or pectin which are highly sensitive to particular salts, e.g., calcium.

2. Polymer Dissolution
a. Methods of Dissolution

Neither magic or miracle are responsible for complete and proper dissolution of the CMC polymer. However, there are times when the frustrated food technologist may believe the above to be true. The physical appearance of CMC (a dry white powder) would superficially lead one to believe that the material will easily enter solution like other food ingredients of similar appearance, i.e., salt or sugar. This is a common error. Except for the appearance similarity, the colloidal nature of CMC makes it quite dissimilar to most aqueous food ingredients in many characteristics, especially in the way it dissolves. Complete solutions of cellulose gum can readily be prepared when the gum concentration is low and vigorous agitation is applied to the system.

When viewed as a mechanical process, the dissolution of CMC may be considered in two phases, *dispersion* and *hydration*. The most difficult step is first. This is because cellulose gum, like other water-soluble polymers, loves nothing more than itself. As the particles hit the water, without good dispersion, they immediately come together forming little balls or clumps with a swollen outer skin of partially hydrated gum. These are the notorious "fish-eyes". Figure 5 depicts an outstanding example of what can happen when improper dispersion technique is employed. Once the clumped state (Figure 5) is attained, the highly swollen outer skin of the clump impairs hydration of the dry interior and drastically reduces the hydration rate of the gum as well as solution viscosity. To overcome this difficulty, the following dispersion techniques are recommended:

1. Direct addition — Add the gum directly to the vortex of a vigorously agitated body of water. In this method, the rate of addition must be slow enough to allow the particles to separate discreetly (like sprinkling salt on a steak — too much would be undesirable), but fast enough so that all of the gum is added before the vortex disappears. Why? Because it is extremely difficult to thicken an already viscous solution of cellulose gum by adding additional powder. Competition between dry and wet polymer for the

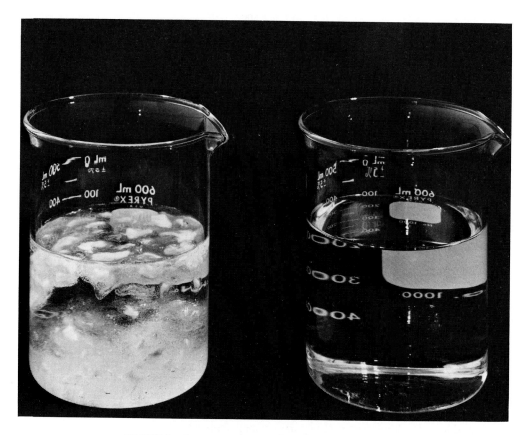

FIGURE 5.   Improper and proper dispersion of cellulose gum.

available water will tend to occur in this situation. This method is most commonly encountered in highly controlled processing situations.

2.   Dry blending — Dispersion of CMC may be accomplished by dry blending the gum with other nonpolymeric ingredients prior to their addition to water. The other particles function to keep the gum particles away from each other. Sugar is the commonest dry diluent used in this method. Generally, ten parts (or more) of sugar to gum is recommended to effectively prevent lumping. A classic example of this dispersion technique is the instant dry mix beverage.

3.   Water miscible nonsolvent — Cellulose gum may be dispersed in a water-miscible nonsolvent such as glycerine, sorbitol, or proplylene glycol which keeps the particles separated. The slurry may then be added to water with agitation to complete the procedure.

An offshoot of this method is to disperse the gum in oil and then add the mixture to an aqueous system with an emulsifier.

Ganz[15] essentially describes this method[3] in much greater detail in a disclosure. Here, the water-miscible nonsolvent may include one or all of the following to optimize the procedure: a compatible salt, a surfactant, and a suspending aid. The salt helps to minimize paste formation and compaction of the water-soluble gum in the nonsolvent when it settles. The surfactant facilitates redispersion of the gum when it settles in the water-miscible nonsolvent. The suspending aid functions just as the name implies to prevent gum sedimentation in the nonsolvent.

4.   The mixing device — Various mechanical devices have been used for placing CMC in solution. One of the most efficient of these units (which is particularly suited to

plant processing conditions) is the use of a special stainless steel mixing device designed by Hercules, Inc. With this device, the gum is fed through a smooth wall funnel into a water jet eductor where it is dispersed by the turbulence of water flowing at high velocity (20 to 40 psi; 10 to 20 gpm). The water flow also creates suction effect which helps the feeding of the gum into the apparatus. Under optimum conditions, the particles are instantly wetted out and the gum from the effluent is 80 to 90% hydrated. Figures 6 and 7 depict the device and its commercial installation.

5. Miscellaneous techniques — Various particle size grades of CMC are available from manufacturers to facilitate dispersion and hydration of the gum in different food applications. Larger or coarse CMC granules are utilized to improve dispersibility when agitation is inadequate to produce a good vortex in the solvent. Ice cream is a good example of a product usually made under poor mixing conditions where a coarse CMC type would be suitable. For food applications, such as dry mix beverages or instant dry mix desserts, needing a rapid solution time, cellulose gum having a fine particle size is the best candidate. However, fine particle size CMC is difficult to disperse (greater surface area and number of particles) and special precautions during dispersion (such as methods 2 and 3) must be heeded.

Improvement in CMC dispersion and concurrent inhibition of the tendency to clump has been accomplished by adding small amounts of dioctyl sodium sulfosuccinate to the gum prior to its dissolution in water.[16] Agglomeration of CMC has also been found to improve both dispersibility and hydration rate of the material.[17]

Rigler et al.[129] describes a unique dispersion technique for CMC whereby the gum may be applied to a food system or solution requiring thickening by admixing with a halogen-donating compound (salt), sodium or potassium bicarbonate, and an organic acid (e.g., citric, adipic, or tartaric). The halogen-contributing compound is believed by the inventors to enhance hydration of the gum. The process resolves gum dispersion problems (no lumps) by utilizing the bursting action that results form the combination of bicarbonate and acid in aqueous medium to yield carbon dioxide. This effervescence creates enough turbulence in the system to disperse the CMC efficiently. A suggested formulation for achieving the overall properties described would include 50 to 75% CMC, 15 to 25% halogen-donating compound, 10 to 15% sodium or potassium bicarbonate and 3 to 10% organic acid.

b. Theory of Polymer Dissolution

Dissolution of the CMC polymer is a rather complex, but interesting chemical-physical phenomenon. As mentioned previously, cellulose is insoluble in water. Numerous strong interpolymer associations (crystalline areas) hold the cellulose chains together and prevent the penetration of water which dissolves the cellulose. When larger carboxymethyl groups replace the hydroxyl groups, the chains are separated and the unreacted hydroxyl groups (normally tied up by hydrogen bonding) become more available for association with water. This is the fundamental explanation as to why cellulose gum is water soluble above a certain minimum DS while the basic cellulose molecule is insoluble in water.

The carboxymethyl group is the functional moiety in the chemical facet of polymer dissolution. It is generally accepted that the chemical charge on the polymer permits initial access into aqueous solution. The solvent attaches itself to the cellulose gum through association with polar or ionic groups on the polymer and, to a lesser extent, through hydrogen bonding. In addition, the solvent may associate with the mobile counter-ions (Na^+) derived from the ionic cellulose gum, thereby reducing the energy required to separate the counter-ion from the vicinity of the polymeric ion and promote dispersion.[21] Complete solubility and subsequent solution properties are then maintained when the chains are held apart by mutual repulsion of like-charged anionic groups (Coulombic repulsion) which allows entrance of water and full interaction with chain hydroxyls. However, some debate exists among

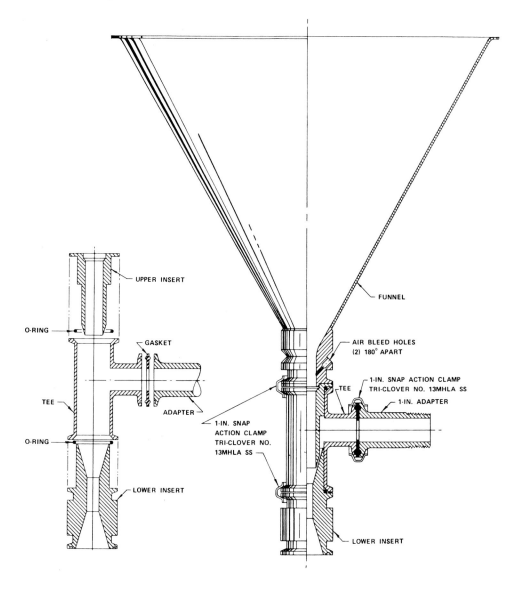

FIGURE 6.   Hercules mixing device.

polymer chemists as to which factor (the polar carboxymethyl moiety or Coulombic repulsion) exerts the greatest importance in the solubilization of CMC.

Concurrent to the above chemical mechanism, a physical process also takes place when CMC undergoes dissolution. Figures 8 and 9 illustrate the idealized disaggregation of CMC as the polymer enters a favorable solvent. State I represents the undisturbed dry polymer in complete aggregation. If the solvent has sufficient solvating power (water is best), the ionic carboxymethyl group interacts with the solvent and helps open the chains. A physical swelling of the polymer chains with solvent (State IA or IB) ensues as well as a viscosity increase (Figure 9) which reflects partial solubilization as chain segments go into solution. State II represents a point of maximum swelling where all the liquid is imbibed into the particles to achieve maximum viscosity. Here the discrete gel particles occupy almost the entire volume of the system. At State II, incomplete disaggregation exists, because some, but not all, of the internal associations are broken. In going from State II to State III, more of the internal associations become broken. The chains become less deformed by swelling (hence the

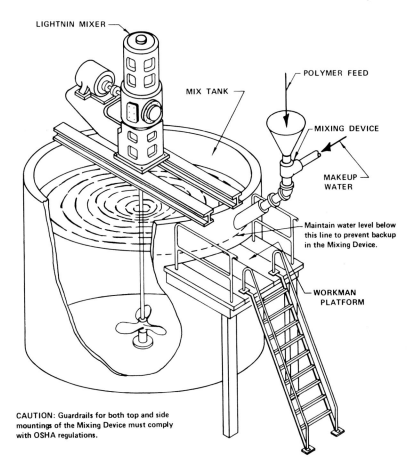

LIGHTNIN MIXER

POLYMER FEED

MIX TANK

MIXING DEVICE

MAKEUP
WATER

Maintain water level below
this line to prevent backup
in the Mixing Device.

WORKMAN
PLATFORM

CAUTION: Guardrails for both top and side
mountings of the Mixing Device must comply
with OSHA regulations.

FIGURE 7.   Typical installation for side mounting the mixing device.

viscosity decrease), but more dispersed. State III represents complete molecular dispersion. The chains are fully disaggregated, an equilibrium viscosity is reached, and dissolution is finalized.

c. Factors Influencing Polymer Dissolution

Various polymer parameters influence the disaggregation curve that describes the dissolution of the CMC polymer. The shape of the curve depends on the type of CMC used, the energy applied to the system, the solvating power of the solvent and the presence of other solvents in the system. Coarse particle size CMC types are more easily dispersed, but require more energy input and time for solvent to penetrate the larger granules and promote disaggregation. Fine particle size CMC types disaggregate and hydrate rapidly with minimal energy input provided that optimum dispersion is employed.

The dispersion and dissolution properties of cellulosics are also determined by the DS and DP that characterize the CMC type. The higher the DS (the average number of polar substituents on the cellulose chain), the more readily the gum dissolves. Increasing DS makes the gum more hydrophilic. A more uniform substitution on the chain enhances disaggregation since fewer crystalline areas (gel aggregates) will exist. The DP or degree or polymerization of the polymer governs the molecular weight of the CMC type. The lower the DP or molecular weight, the shorter the chain, hence the faster rate of dissolution.

The effect of solvent strength on disaggregation is shown in Figure 10. As the solvating power increases (goes from a lean nonsolvent to a polar aqueous solvent), disaggregation

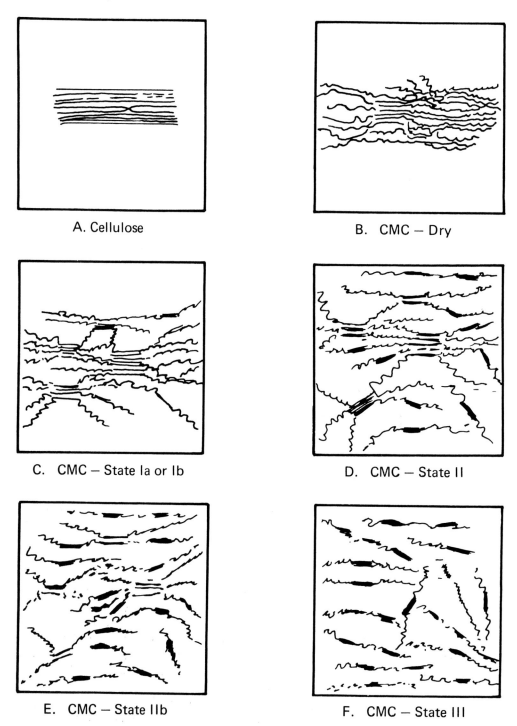

A. Cellulose

B. CMC — Dry

C. CMC — State Ia or Ib

D. CMC — State II

E. CMC — State IIb

F. CMC — State III

FIGURE 8.   Cellulose gum in various states of aggregation.

is promoted. Note the similarity in shape of the solvent strength curves to the idealized disaggregation curve in Figure 9. The mutual effect of increasing DS and solvating power on dissolution is also shown in Figure 10.

Given the most optimum solvation, the temperature of the solvent system plays a further role. Generally, as solvent temperature increases, the disaggregation process is accelerated.

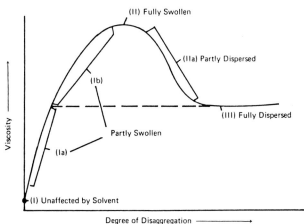

FIGURE 9. Effect of polymer disaggregation on viscosity of cellulose gum.

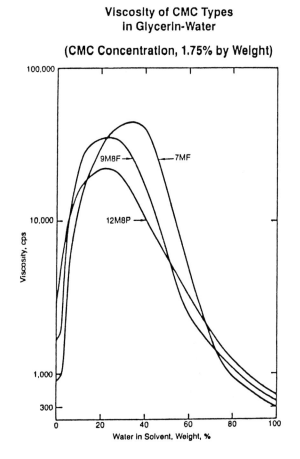

FIGURE 10. Effect of solvent strength on disaggregation of cellulose gum.

However, increasing solvent temperature enhances the affinity of the particles for each other (fisheye formation). This may become a problem when dispersion is inadequate.

The influence of solutes (salts) on the dissolution of CMC is illustrated in Figure 11.

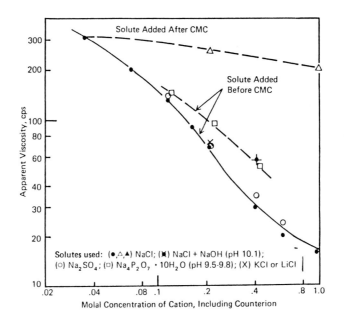

FIGURE 11.   Effect of solutes on the viscosity of cellulose gum solutions.

Salts tend to decrease the hydration of the gum and the corresponding solution viscosity. Salt cations provide a screen of counter-ions around the carboxyl group, thereby leading to a reduction of the repulsive forces between carboxyl groups which disfavors dissolution.[19]

Order of addition of solutes and hydrocolloid regulates the interaction. When salt is added to fully dissolved CMC, the chains are already separated and viscosity depression by the salt is minimal (Figure 11). However, when CMC is added to saline solution (Figure 11) the chains never have a chance to separate by mutual repulsion. Consequently, dissolution is inhibited and viscosity development is markedly reduced.

Figuratively speaking, salts function to maintain the crystalline regions along the chain, especially when order of addition is wrong. Therefore, it is no surprise that the effect of solutes on CMC dissolution is less pronounced with high DS or uniformly substituted material which contains fewer crystalline areas.

B. Rheological Behavior

Sodium carboxymethylcellulose exhibits several interesting and useful rheological properties upon reaching complete dissolution. To the food technologist, knowledge of these rheological properties can serve as a guide for selection of the proper cellulose gum type to achieve desired textural characteristics in the finished food system.

The single most important rheological property of cellulose gum is the ability to impart viscosity to aqueous systems, foods, or beverages.

Viscosity is a measure of system resistance to flow when subjected to an applied shearing force. In simple solutions where the dissolved material is low in molecular weight, is nonassociating, and limited solute-solvent interactions occur, the flow is directly proportional to the force applied. The system is said to be *Newtonian* and the viscosity remains constant as shear stress varies. More complex solutions, however, like CMC respond in a nonlinear manner to applied stress. Here, the dissolved molecules are large, the tendency to reassociate is high and the solvent must exert some solvating force to maintain the polymer in solution. Such solutions are classified as *non-Newtonian*, and the viscosity changes as shearing action is varied.

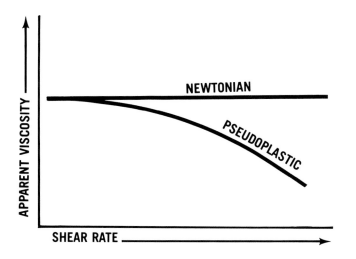

FIGURE 12. Pseudoplastic flow.

In common with other linear water-soluble polymers, cellulose gum solutions are pseudoplastic; that is, the viscosity decreases as the rate of shear increases (Figure 12). However, if the applied force is removed, the solution will revert instantly to its nonsheared rheology. As a practical food example, cake frostings stabilized with high-viscosity cellulose gum are quite immobile when ''poured'' from their container (low shear), but thin substantially when spread onto cakes with a knife (higher shear).[49] Upon cessation of spreading (shear removal), the frosting is again immobile.

One may contemplate as to why cellulose gum in solution exhibits pseudoplasticity. Basically, this rheological property occurs when the long-chain molecules orient themselves in the direction of flow. Figure 13 provides an oversimplification of the phenomenon involving a Brookfield rotational viscometer. At rest or low shear, CMC polymer chains exist in a random state. At higher shear, the random chains elongate and orient themselves to the direction of flow triggered by the applied force from the faster turning spindle. As chain alignment improves, the spindle turns more easily in the matrix, because the resistance to flow — the viscosity — decreases. When the instrument is turned off, the random state is instantly reassumed.

The pseudoplastic behavior of CMC varies with the molecular weight (chain length) and DS of the particular CMC grade. In Figure 14, the effect of increasing shear on different molecular weight grades of CMC may be observed. All behave as Newtonian materials at very low shear rates. As the shear rate increases, low molecular weight types become less pseudoplastic than high molecular weight types.

Thixotropic flow is another rheological property characteristic of cellulose gum. The term itself, *thixotropic,* is derived from the Greek — ''thixis'' meaning to strike or touch and ''tropo'' meaning to turn or change. The Greek nomenclature provides an insight as to the nature of this rheological behavior.

Basically, thixotropy is a type of pseudoplastic flow with a time dependency. Figure 15 illustrates the phenomenon. Similar to pseudoplastic flow, thixotropic cellulose gum solutions undergo shear thinning with an increase in applied shear force. However, when the shear force is removed, time is required for the thixotropic solution to revert to its original viscosity, whereas a pseudoplastic solution reverts back instantly after shear removal. Upon preparation, thixotropic cellulose gum solutions or products containing the gum will display an increase in apparent viscosity while remaining at rest for a protracted period of time. If sufficient agitation (a shear force) is applied, the viscosity will be reduced. If a rest period is renewed, the viscosity will again begin to build.

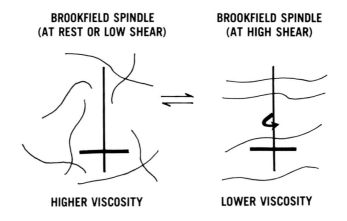

**BROOKFIELD SPINDLE
(AT REST OR LOW SHEAR)**

**BROOKFIELD SPINDLE
(AT HIGH SHEAR)**

HIGHER VISCOSITY

LOWER VISCOSITY

FIGURE 13.   Why cellulose gum solutions are pseudoplastic.

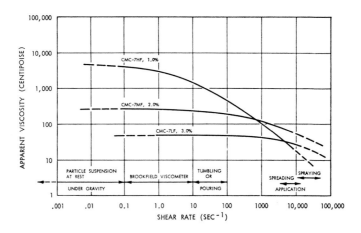

FIGURE 14.   Typical flow behavior of cellulose gum solutions.

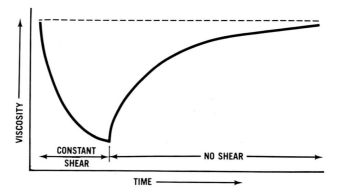

FIGURE 15.   Ideal curve for thixotropic solutions.

In extreme cases, the thixotropic viscosity build may approach a gel-like consistency.[18,19,23] If enough force is applied, the structure is broken and consistency is lessened. Such extreme thixotropic behavior is analogous (but not identical) to the rheology of a modified Bingham body where a yield stress is required to initiate flow.[22]

FIGURE 16. Thixotropic and nonthixotropic solutions of cellulose gum. Solution of regular Hercules cellulose gum (left) is thixotropic; "S"-type Hercules cellulose gum (right) is essentially nonthixotropic.

The key factor responsible for the thixotropic nature of certain cellulose gum solutions is the uniformity of substitution and/or lack of substituents which cause "insoluble" (somewhat hydrophobic) regions along the chain.[18,22] These insoluble regions or "gel centers" tend to reassociate with time, forming a three-dimensional network or structure which translates as a viscosity increase. Shear, of course, will break the structure, but not permanently. Elliott and Ganz[23] concluded that thixotropy in CMC solutions arises because of the presence of unsubstituted crystalline residues in the CMC. These would be present as fringe micelles which could form cross-linking centers that would entrap a relatively large amount of molecularly dispersed CMC by electrostatic hydrogen bonding, or Van der Waals forces, and thus enable a three-dimensional structure to be set up.[23]

The thixotropy phenomenon is concentration dependent; with more gum in solution, a "crowding effect" occurs which enhances the magnitude of the thixotropic increase. High viscosity types as well as low DS types of CMC (0.4 to 0.7) will generally display thixotropy, because these species have the greatest amount of unsubstituted or insoluble regions.

Figure 16 visualizes the difference between a uniformly substituted, smooth flowing, nonthixotropic CMC solution vs. a randomly substituted, thixotropic solution of CMC with structured flow. Solution appearance can be altered from a thixotropic (applesauce consistency) to a very smooth (syruplike) consistency with no change in DS by special reaction conditions and raw material selection during manufacture.[6]

Uniformly substituted, smooth flowing cellulose gum is highly desirable as a texturizer for food systems such as syrups, puddings, or frostings where smooth consistency is a must. Thixotropic cellulose gum finds use in foods requiring a structure "grainy" texture. Sauces and purées are typical examples.

1. Gelation

Sodium carboxymethylcellulose will undergo gelation in the presence of specific cations or shear conditions.

Trivalent cations, such as Al^{+3}, Cr^{+3}, or Fe^{+3}, may precipitate, form very thixotropic solutions, or gel CMC. Careful selection of ion concentration, regulation of pH and use of a chelating agent for controlled ion release favor gelation. Depending on conditions and technique, various gel strengths may be produced. Ganz[18] suggests that gel formation rather than precipitation may be achieved by avoiding localized concentrations of trivalent cation through the use of sparingly soluble salts such as basic aluminum acetate or by the slow addition of a dilute solution of a more soluble salt, e.g., alum. As to the mechanism of gelation of cellulose gum with aluminum or other polyvalent ions, it is believed that the ion serves as a cross-linking agent for polymer chains through salt formation with adjacent anionic groups.[18] High DS cellulose gum is more apt to gel with trivalent cations. Unfortunately, aluminum cellulose gum gels have not found an application in the food industry due to their astringent taste and poor mouthfeel.

Gels may be prepared from cellulose gum when polymer of sufficiently low DS is subjected to high shear conditions.[19,22,23] Elliott and Ganz[22] demonstrated this gelation technique by preparing gels from low-to-high DS CMC (5% solutions) at high shear in a Waring Blendor®. Subsequent characterization of the effect of DS on gel strength using a Weissenberg Rheogoniometer was made. The sample made from the lowest DS type (0.18) showed the sharpest stress peak and extremely rapid stress decay on the Rheogoniometer trace of all trials, indicating firmest gel strength.[22] These gels were judged organoleptically to be unctuous, whereas those of higher DS were not. Matz[28] defined unctuous foods as fatty substances. *Webster's Third International Dictionary* defines "unctuous" as having the nature or qualities of unguent or ointment, or smooth and greasy in texture.

Elliot and Ganz postulated the mechanism of CMC shear-gels as follows: "The formation of a gel when a thixotropic CMC solution without cross-linking cations is subjected to high power input stirring is believed to arise because of the dispersion and disaggregation of the fringe micelles arising from the crystalline residues, thus providing more potential cross-linking points, and not by disruption of the individual cellulose crystalline residues."[11,23,29] One would anticipate more crystalline residues to be present in low DS CMC than in high DS CMC.

Nijhoff [30] has been granted a U.S. patent on the use of low DS CMC to form unctuous gels for low-calorie spreads, dressings, and desserts processed with homogenization shear. This work cited CMC in a DS range of 0.35 to 0.40 as optimum for the desired effect.

C. Solution Viscosity
1. Effect of Concentration

Figure 17 shows the relationship of concentration and apparent viscosity of various cellulose gum types available from one manufacturer. The increase in viscosity is not directly proportional to changes in concentration, but instead an exponential function exists. A good rule of thumb is that doubling the concentration will usually increase the viscosity tenfold. It is particularly important to remember the effect of concentration on viscosity when developing a food product or making minor adjustments to products during processing.

At a fixed concentration, the DP (degree of polymerization) regulates the viscosity of cellulose gum. This relationship is straightforward — the higher the DP (the longer the polymer chain), the higher the viscosity of the CMC derived from it.

Food grade cellulose gum has a molecular weight range between 40,000 and 1,000,000.[6]

2. Effect of Temperature

As with most water-soluble polymers, the viscosity of cellulose gum solutions decreases with increasing temperature (Figure 18). Under normal conditions, the effect of temperature on viscosity is reversible and raising or lowering the solution temperature has no permanent effect on the viscosity characteristics of the solution. However, prolonged heating at very high temperatures will tend to depolymerize cellulose gum and permanently decrease the apparent viscosity. Hence cellulose gum, like many other long chain polymers, is not perfectly retort stable.

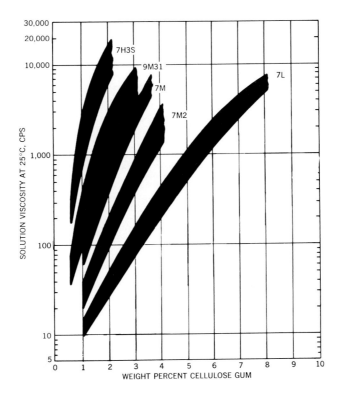

FIGURE 17. Effect of concentration on viscosity — aqueous solutions of Hercules cellulose gum. Bands indicate the viscosity range (for concentrations up to 10%).

3. Effect of pH

The apparent viscosity of cellulose gum solutions is relatively stable over a wide range of pH of about 4 to 10 (Figure 19). Optimum stability of the gum occurs near neutral pH. Most foods are found in the 4 to 7 pH range.

As indicated previously, at about pH 3.0, precipitation of the free acid form of cellulose gum takes place. Consequently, an upswing of the curve in Figure 19 results.

At extreme alkaline pH (above 10), a decrease in viscosity is observed. Since alkaline foods are rare, this is generally not a problem.

In some food systems, cellulose gum is subjected to conditions which may cause it to undergo acid-catalyzed hydrolysis, resulting in permanent loss of viscosity. The reaction takes place at the ether linkage between two unsubstituted anhydroglucose units (the so-called doublets) and is accelerated by increasing the temperature or further lowering of pH. Since the chain doublets or unsubstituted regions are greatly reduced with high DS or uniformly substituted CMC, these types of cellulose gum are preferable when acidic conditions prevail.

4. Effect of Electrolytes

Previously, the effect of electrolytes (salts) upon dissolution and gelation of CMC was discussed. In general, the effect of salts on cellulose gum solution viscosity varies with the particular salt, its concentration, pH of the solution, the magnitude and uniformity of the degree of substitution of the cellulose gum, and the order of addition.

Monovalent (alkali metal) cations interact with carboxymethylcellulose to form soluble salts. In aqueous systems containing these cations, cellulose gum is compatible and full

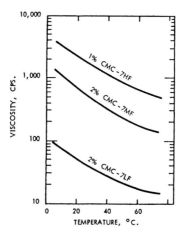

FIGURE 18.   Effect of temperature on the viscosity of Hercules cellulose gum solutions.

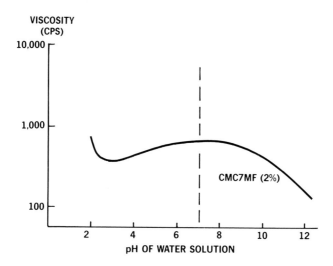

FIGURE 19.   Effect of pH on the viscosity of Hercules cellulose gum solutions.

viscosity development depends primarily on the order of addition. High DS cellulose gum is best for situations requiring good salt tolerance.

Divalent cations are borderline in behavior with cellulose gum. Calcium ion, for example, if present in sufficient concentration, such as in hard water, prevents cellulose gum from developing peak viscosity, and these dispersions may be hazy. At much higher concentrations, calcium ions precipitate cellulose gum from solution. Magnesium and ferrous ions exert a similar effect on cellulose gum. Uniformly substituted cellulose gum is preferred when moderate concentrations of divalent cation are present.

Trivalent cations will precipitate, form hazy dispersions, or as previously mentioned, gel cellulose gum. Specific factors within the system will determine which of these interactions is most likely to occur.

D. Compatibility

Sodium carboxymethylcellulose is compatible in solution with most common food ingredients including protein, sugar, starches, and other hydrocolloids.

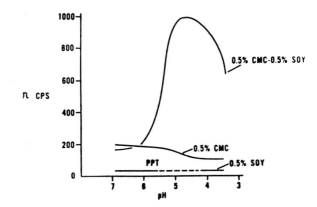

FIGURE 20. Effect of pH on interaction of cellulose gum type 7H3SF with soy protein.

1. Protein Reactivity

Proteins are important ingredients found in many food systems. However, their use has limitations. Some proteins are insoluble in aqueous solution at the isoelectric pH region and most undergo denaturation upon acidification or by heating at elevated temperatures.

When cellulose gum is added to systems containing protein, unexpected textural effects and changes or improvements in the functional properties of the protein can result. Interaction of protein and cellulose gum usually results in the formation of a soluble complex or an insoluble complex (precipitate). The nature and extent of the interaction may be observed by viscosity measurement. As proteins differ widely in their properties, information gained from one system generally cannot be transposed *in toto* to another.

The reaction between water-soluble proteins and the anionic polymer, CMC, is largely an ionic mechanism although other contributing factors such as hydrogen bonding, van der Waals forces and steric effects are undoubtedly involved.[18,19,31,32] Further evidence that the CMC-protein interaction occurs ionically is provided by nonionic gums which show relatively little interaction with protein.[18,31]

Figure 20 illustrates the interaction of high-viscosity, uniformly substituted cellulose gum with soy protein. In the isoelectric region of pH 4.5 to 5.0, normally insoluble soy protein is solubilized by an equal amount of cellulose gum. This is shown by the large increase in the apparent viscosity of the mixture. Interestingly enough, the soy-CMC complex has improved stability (no precipitation or viscosity change) upon heating and cooling.[18,31,33]

Other factors, including molecular weight of the interacting components and the presence of salts, influence the complex formation shown in Figure 20. If the cellulose gum chain length is too short (a low molecular weight type), no solubilization of the soy protein occurs. Salts tend to depress the viscosity of the mixture, but do not inhibit complex formation entirely.[31]

The ionic mechanism of the soy-CMC interaction may be described using Figure 20. In the isoelectric region, the negatively charged cellulose gum combines with positively charged groups of the protein, and the complex formed is believed to be more polar and therefore soluble.[19] Upon addition of alkali (higher pH), the positively charged amine group is neutralized, and combination with anionic CMC is inhibited.[18,31,33] As this occurs, the viscosity is lowered, approaching that normally expected from a blend. Upon acidification below the isoelectric pH, protonation of free carboxyl groups occurs and a precipitate may form.[18]

Interaction similar to that described in Figure 20 can occur with CMC and other proteins such as gelatin or casein. Ganz[18] showed that significant viscosity elevation occurs when specific gelatin types are combined with CMC-7H3SF at varying concentration and pH. Thompson[34] demonstrated that complexes having unique chemical and physical properties can be prepared by reacting gelatin and CMC under controlled conditions. An interesting reaction occurs when gelatin, which is soluble in hot water only, is dried with cellulose gum to yield a product that is cold-water-soluble.[46]

Various workers have utilized the protein reactivity of cellulose gum for isolation, purification, and/or recovery of either casein or whey from commercial processes or products.[35-40]

CMC-protein complex formation becomes important in the preparation and stabilization of fruit-flavored milk beverages or directly acidified sour milk products where the acidic environment would normally precipitate the milk protein.[19,41,42] The CMC-protein interaction creates a protective colloid effect.

Ganz[43] showed that liquid egg white dried with sodium carboxymethylcellulose had superior characteristics for use in baked products (whipping, binding, and stabilization) compared to a dry mix of the two ingredients.

2. Interaction with Sugars

Cellulose gum modifies the water behavior of sugar solutions in several capacities. Combinations of cellulose gum and sugar exhibit a progressive, marked "boost" in viscosity with increasing gum concentration.[18,19] The effect, which is much greater than that anticipated from the component viscosities, is believed to be the result of a "crowding" mechanism. However, the possibility of some H-bonding between the sugar and the gum cannot be excluded. Sugar is also known to enhance the magnitude of thixotropic viscosity build characteristic of randomly substituted cellulose gum.

Cellulose gum decreases syneresis in foods high in sugar content by binding up free water in the system. Control of sugar crystal size and rate of sugar crystallization in concentrated sugar systems has also been shown by the incorporation of small amounts of cellulose gum.[44] Low-viscosity grades of cellulose gum are most effective as crystallization inhibitors, whereas high-viscosity grades are most effective in prevention of water loss from sugar solutions.

Such functionality has important implications in the stabilization of confectionery and bakery products (fillings) where desired plasticity, gloss, and nongritty mouthfeel are required.

3. Interaction with Starches

Cellulose gum is used in combination with starch in many food systems, and compatibility between the two ingredients is generally observed.

Ganz[18] showed that cellulose gum in combination with wheat starch lowered the gelatinization temperature, increased peak viscosity, and decreased cookout.

Combinations of cellulose gum and starches are routinely employed in specific foods (desserts) where they modify each other's textural contribution or optimize texture. Such interactions are currently under investigation by several ingredient suppliers.

4. Compatibility with Other Hydrocolloids

Cellulose gum is compatible with most other water-soluble polymers over a wide range of concentration. When anionic cellulose gum and nonionic water soluble gums, such as hydroxypropylcellulose (Klucel®), locust bean gum, or guar gum, are admixed in aqueous solution, the compatibility is such that a marked synergistic viscosity increase results.[20] This synergistic effect occurs most significantly when ionic and nonionic components possess large regions of unsubstitution along the chain lengths. Thus locust bean gum (with its smooth region) undergoes greater synergistic viscosity build with cellulose gum compared to more uniformly substituted guar gum of similar viscosity.[45]

Cellulose gum has not been shown to undergo any significant synergistic behavior with other anionic gums, but they are compatible in food systems. For example, ice cream stabilizer blends routinely employ cellulose gum and carrageenan together.

Cellulose gum has been used to modify carrageenan gel texture and help prevent water exudation from brittle kappa-carrageenan gels.

E. Film Formation

CMC, like other water-soluble gums derived from cellulose, will form films. By casting and evaporating the water from cellulose gum solutions, clear films may be obtained. Better strength and flexibility of such films result when they are prepared with high-viscosity-type CMC.[6] Incorporation of plasticizing agents, such as glycerol, can improve the flexibility and elongation (stretchability) of cellulose gum films.

Fats and oils, grease, and organic solvents do not affect cellulose gum films. This property makes the CMC film functional in food applications requiring prevention of fat absorption or a barrier to organic agents.

Obviously, cellulose gum films are water soluble. Dry CMC films may be insolubilized by treatment with aqueous solution containing aluminum cation which exchanges with the sodium ion.[6,9]

F. Stability of Cellulose Gum

1. Microbial Stability

Cellulose gum in solution will remain stable at room temperature for a prolonged period of time with no loss in viscosity if free from mold, fungi, or bacterial growth. Typical heating processes, such as 30 min at 80°C or 1 min at 100°C, are generally sufficient for prevention of microbiological degradation of cellulose gum in solution or in food products without affecting the properties of the gum. For long-term storage of solutions or foods containing the gum, a food-approved preservative should be included as an additional safeguard to prevent viscosity loss. Commercially available preservatives such as sodium benzoate, sorbic acid and its salts, or sodium propionate are commonly utilized to protect the gum.

Cellulose gum in solution is particularly susceptible to degradation by active cellulases. These enzymes are generated by airborne bacteria and are present in certain food ingredients such as vegetable juices, onion, garlic, etc. Cellulases can reduce viscous cellulose gum solutions to a water-like consistency in a few hours. Inactivation of cellulases may be accomplished by heat treatment or adjustment away from the pH optimum for enzyme activity.

Cellulose gum is sometimes incompatible with microbial gums when combinations of such are utilized in environments where cellulases from the microbial gum remain active.

Biological attack of cellulose gum occurs predominately in solution systems and is quite rare with the dry form of the gum, providing that moisture content is kept to a minimum. Generally speaking, the product as produced and packaged is relatively free of harmful microbes. The gum, when made from manufacturing systems using solvents, is essentially aseptic.[6] Most microbiological organisms which cause degradation of cellulose gum originate from outside sources such as food processing equipment or influent water. In food processing plants, where mixing tanks, piping, pumps, valves, etc. are not thoroughly cleaned after each use, conditions would be ideal for microbial growth to harm products stabilized with cellulose gum. Proper sanitary procedures are essential to the quality of the food product, the safety of the consumer, and the stability of cellulose gum.

2. Other Stability Factors

Sunlight (UV) and entrained air are causative agents which influence the degradation of cellulose gum. The presence of molecular oxygen in solutions or foods containing cellulose gum may result in degradation of the polymer by a reaction process similar to autoxidation of lipids. Heavy metals can serve as catalysts in the oxidative degradation of cellulose gum.

Therefore, chelators, such as sodium citrate or sodium hexametaphosphate, will help stabilize cellulose gum (indirectly) in food systems where entrained air and heavy metals exist.

G. Physical Data

Typical physical properties and particle size designations for Hercules cellulose gum are summarized in Table 5 and 6, respectively.

Another physical property characteristic of cellulose gum is that the material is quite hygroscopic. This feature of the gum is both functional and detrimental. Adequate packaging and handling of CMC is essential for manufacturers (both producers and users of the gum) since the hydrocolloid will readily "pick up" moisture from the atmosphere and become unusable if exposure to humidity persists for significant time periods.

Conversely, the "attractiveness" of water to cellulose gum is quite useful in foods where the gum serves as an excellent water binder. Figure 21 shows the effect of increasing DS on equilibrium moisture content of the material as realtive humidity increases. Obviously, higher DS grades of the gum function as the best water binders. For this reason, high DS types of cellulose gum are preferred in foods requiring moisture retention for maintenance of eating qualities. Cakes and the frostings that coat them are primary examples of products that rely on the hygroscopicity of cellulose gum.

V. FOOD APPLICATIONS

Since the introduction of sodium carboxymethylcellulose to the food industry in the late 1940s as a substitute for gelatin in ice cream, use of the gum as a functional ingredient has grown and diversified in a myriad of food applications. Table 7 provides a list of known food applications for cellulose gum. This compilation is by no means complete and probably is already outdated as new food uses for cellulose gum continue to be discovered from within R & D departments throughout the food industry.

A. Dairy Food Applications
1. Ice Cream Stabilization

Ice cream is perhaps the best known and most famous dessert entrée. Its popularity with children is unsurpassed. Even so, all age generations find delight in this frozen delicacy. The appeal of ice cream to folks of all ages arises from the numerous organoleptic attributes which characterize the frozen dessert. As a food, ice cream is wholesome, highly nutritious, cool and refreshing, pleasing to the palate, prepared in a variety of flavors and colors, sweet, creamy, smooth, tantalizing, and affordable.

Historically, ice cream is an ancient food substance. The first prototype is reported to have originated during the reign of the mad Roman emperor, Nero, in the first century, about 62 A.D.[10] It seems that the emperor was addicted not only to fiddling, but also to eating a concoction made up of snow, nectar, fruit pulps, and honey.[10] Since refrigeration was not yet invented, Nero employed fast runners to bring back snow from the Apennine Mountains to prepare this delicacy. In the 13th century, Marco Polo returned home to Venice from his famous journey to the Orient bringing with him riches, substances, and knowledge unknown to the European continent. Among these discoveries were recipes for water ices said to have been used in Asia for thousands of years. Thereafter, knowledge of ice cream and its preparation spread slowly throughout Europe.

Ice cream was probably brought to America by early colonial settlers. A letter written in 1700 by a guest of Governor Bladen of Maryland described having been served ice cream.[50] President Washington, the father of our country, was a tower of strength, yet among his few weaknesses was ice cream, of which he was very fond. Washington provided for the inclusion of rudimentary ice cream making equipment to the culinary quarters of his beloved

Table 5
TYPICAL PROPERTIES OF HERCULES
CELLULOSE GUM

Polymer — as shipped
  Sodium carboxymethylcellulose — dry basis, %   99.5 min
  Moisture content, max, %   8.0
  Browning temperature, °C   227
  Charring temperature, °C   252
  Bulk density, g/mℓ 0.75
  Biological oxygen demand (BOD)[a]
    7H Type   11,000 ppm
    7L Type   17,300 ppm
Solutions
  Specific gravity, 2% solution, 25°C   1.0068
  Refractive index, 2% solution, 25°C   1.3355
  pH, 2% solution   7.5
  Surface tension, 1% solution, dynes/cm at 25°C   71
  Bulking value in solution, gal/lb.   0.0652
Typical films (air-dried)
  Density, g/mℓ 1.59
  Refractive index   1.515

[a] After 5 days' incubation. Under these conditions cornstarch has a BOD of over 800,000 ppm.

Table 6
PARTICLE SIZE DESIGNATIONS OF
HERCULES CELLULOSE GUM

Designation	Description	Particle size[a]
None	Regular	Maximum 1% retained on 30 mesh; maximum 5% retained on 40 mesh (10% for certain types)
C	Coarse	Maximum 1% retained on 20 mesh; maximum 50% through 40 mesh; maximum 5% through 80 mesh
X	Fine	Maximum 0.5% retained on 60 mesh; minimum 80% through 200 mesh

[a] All screens are U.S. Bureau of Standards sieve series.

estate, Mt. Vernon. Ice cream was served in the White House at the second inaugural ball of President Madison in 1812 by Dolly Madison. Today, this history is reflected in the fact that a leading brand of ice cream bears the name of our early First Lady.

In 1851, the birthplace of the American ice cream industry was established at Baltimore, Maryland when Jacob Fussell opened the first wholesale ice cream factory at Hillen and Exeter Streets.[50]

Arbuckle[50] states that excellent ice cream can be made and considerable amounts are made without the use of stabilizer or emulsifier. It seems that natural substances (milk protein, phosphates, lecithin, etc.) contained in traditional ice cream ingredients impart some stability and emulsification. Although such is perhaps true for those ice creams claiming "all-natural" status or homemade ice cream, by and large, the bulk of commercial ice creams employ

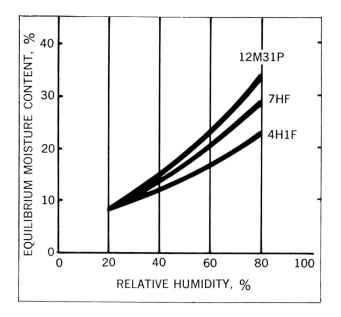

FIGURE 21.   Effect of relative humidity on equilibrium moisture content of Hercules cellulose gum at 25°C.

one or more stabilizers and/or emulsifiers to impart and, more importantly, maintain added superior eating qualities which are not possible in unstabilized ice cream. Stabilization of ice cream with hydrocolloids, either the liquid mix prior to processing or the finished product, has facilitated vast monetary savings for manufacturers because of better uniformity, extended shelf life, improved tolerance to adverse temperature conditions and a more exact control of the product during the various manufacturing processes.

It may be of interest to food historians that in 1915, J.H. Frandsen, a pioneer in the ice cream industry, coined the work "stabilizer" to designate a group of substances used in ice cream which at that time were referred to as "holders", "colloids", "binders", and "fillers".[51] Many workers have reviewed and reported on the functionality and need of stabilizers in ice cream and related frozen desserts.[10,50,51-57,59-60,64]

Potter and Williams[52] listed the following important factors to be considered in selecting a stabilizer: (1) ease of incorporation into the mix, (2) effect on the viscosity and whipping properties of the mix, (3) type of body produced in the ice cream, (4) effect on meltdown characteristics of the ice cream, (5) ability of the stabilizer to retard ice crystal growth, (6) quantity required to produce the desired stabilization, and (7) cost.[52]

Moss[55] compared stabilized to unstabilized ice cream and reported the former had a heavier body, did not taste as cold, and melted down to a creamier consistency. Doane and Keeney[60] explained that in many food applications, stabilizers increase the viscosity of the unfrozen portion which restricts molecule migration to crystal nuclei, thereby limiting crystal size. Also, the ability to bind or hold large amounts of moisture maintains a smooth texture. Shipe et al.[61] found that the effect of stabilizers on the freezing characteristics of ice cream were associated with changes in viscosity and the rate of migration of solutes through dialyzing membranes. Solute migration was believed to influence the rate of ice crystal formation in the ice cream. Glicksman[10] stated that it is generally agreed that the basic role of hydrophilic gums in stabilized products is to reduce the amount of free water in the mix by binding it as water of hydration or by immobilizing it within a gel network. He summarized that small percentages of gums in ice cream function to produce good body or "chewiness," smooth texture, uniformity of product, resistance to melting or slow meltdown, control or reduction of ice crystal growth during storage, and freeze/thaw or heat shock resistance.

Table 7
FOOD APPLICATIONS OF SODIUM CARBOXYMETHYLCELLULOSE (CELLULOSE GUM)

Alcoholic beverages	Hot cereal
Aerosol toppings	Ice cream, sherbet, water ices
Baker goods	Ice cream toppings and ripples
Baker's jellies	Ice cream mixes (instant)
Battered and breaded foods	Icings, frostings, glazes, meringues
Beer foam stabilizer	Icing mixes
Breakfast drinks	Imitation jams and jellies
Cake mixes	International foods
Cake toppings	Meat pies
Canned fruits and preserves	Meat sauces and gravies
Cheese spreads	Meringues
Chiffons	Milk beverages
Citrus concentrates	Pancake mixes
Coffee whiteners	Pet foods
Confectionery	Pie and pastry fillings
Dehydrated foods	Preservative coatings
Dietetic foods	Puddings
Doughnuts	Potato salad
Dry mixes (cakes, etc.)	Relishes and condiments
Dry shortenings	Reduced calorie products
Extruded foods	Salad dressings
Fish preservation	Soft drinks and concentrates
Food emulsions	Still beverages
Fountain syrups	Table syrups
Frozen foods	Tacos and tortillas
Gravies and sauces	Whipped topping mixes

Another problem area with ice cream is lactose crystallization or sandiness. Nickerson[62] concluded that the primary factor responsible for the reduction of sandiness was the use of CMC or other vegetable gums which inhibited the formation of nuclei and thereby inhibited growth of undesirable large lactose crystals. Nickerson found in a study of 36 commercial samples held at 12°F for 7 months that neither partial substitution of corn syrup (a doctor) for sugar nor emulsifiers were as effective as hydrocolloids in preventing sandiness.

The key functionality of free water control by gums in ice cream as stated earlier by Glicksman (1969) is further reflected by more recent workers. Cottrell et al.[63] specified that an optimum, gum-derived viscosity is necessary for best final properties in an ice cream mix.

Moore and Schoemaker[59] showed, by graphic rating and time-intensity tests, significant correlations between CMC concentration and the sensory textural properties of coldness, iciness, viscosity, and firmness in vanilla ice cream.

Since the role of hydrocolloids as functional agents in ice cream is now established, it is perhaps no great revelation that ice cream has served as one of the most well-known test vehicles for food hydrocolloids in their history. There is not a gum "afloat" in the food industry that at one time or another has not had its functionality explored in ice cream.

Arbuckle[50] has listed those gums which are permitted or used in ice cream to include: "agar-agar; algin (sodium alginate); propylene glycol alginate; gelatin; gum acacia; guar seed gum; gum karaya; locust bean gum; oat gum; gum tragacanth; carrageenan; salts of carrageenan; furcelleran; salts of furcellaran; lecithin; pectin; psyllium seed husk; and CMC (sodium carboxymethylcellulose)."[50] Among these gums, Arbuckle particularly mentions algin derivatives and CMC as having gained important places as basic stabilizing materials for ice cream.

The introduction of CMC to the food industry (ice cream) as a replacement for gelatin just after World War II was mentioned earlier. In 1945, it was the initial evaluation and subsequent favorable report of Josephson and Dahle that established CMC as an effective ice cream stabilizer.[64] Pompa[65] conducted a similar investigation which further confirmed the functionality of CMC in various ice cream mixes. Burt[66] and Werbin[67] were first to observe that CMC enhanced the whipping properties (aeration) of ice cream mixes.

Although CMC has found utility in many food systems, Glicksman[68] commented that the major application of CMC has been and still is as an ice cream stabilizer. Klose and Glicksman[80] concluded, after a review of ice cream stabilizers, that because of price or functionality, alginates, CMC, and guar or locust bean gums have replaced most other hydrocolloids as functional agents in ice cream. At the present time, after extensive exploration of all existing food gums in ice cream, CMC remains as one of the leading stabilizers for frozen dairy desserts.

The amount and type of CMC required in ice cream will vary with the butterfat and total solids content, the texture desired, and the processing conditions in the ice cream plant. Generally, medium- to high-viscosity grades of CMC are preferred for ice cream as lower molecular weight types lack sufficient viscosity impact needed to achieve proper stabilization and texture improvement. Usually a working range of CMC is recommended for commercial ice cream mixes. Shown below is a simplified recipe which outlines the solids typically found in an ice cream mix.

Butterfat	12.0%
Milk solids (not fat)	11.5%
Sugar	15.0%
Commercial stabilizer	0.25 to 0.4%
(approximately 50% cellulose gum)	

For high solids mixes, as little as 0.12 to 0.15% CMC has proven satisfactory. Mixes low in butterfat, i.e., those just meeting the Standard of Identity of 10%, employ more CMC to maintain desired textural characteristics compared to premium mixes high in butterfat. In chocolate ice cream mixes, it is recommended that approximately a 25% reduction in CMC usage be made compared to amounts added to other varieties of ice cream. Otherwise the mix may be too viscous and too smooth (nonchewy) in texture. Cocoa contains a small percentage of natural gums and mucilages, fat and emulsifiers which helps explain why less CMC is required to stabilize chocolate ice cream.

Dahle[69] suggested a usage of 0.15 to 0.18% CMC in a study of various ice cream stabilizers. Moore and Shoemaker[59] found concentrations of CMC from 0.1 to 0.2% preferable for better firmness, slower meltdown, less iciness, and greater mix viscosity in vanilla ice cream compared to no or lower usages of CMC. CMC type 7HF was used in this study.

Freezing, hardening, and storage conditions in the ice cream plant are important from the standpoint of texture. When less than optimum conditions are encountered, more CMC should be used.

Substitution of corn syrup solids for some of the cane sugar in an ice cream mix to produce additional body is a common practice. When this procedure is followed, a slight reduction in CMC usage will optimize the formulation.

Glicksman[68] stated that CMC met with immediate acceptance upon its introduction as an ice cream stabilizer, but that results are best when CMC was used together with one or more stabilizers, such as carrageenan, locust bean gum, or gelatin. Rothwell and Palmer[70] remarked that the trend in ice cream stabilizers was to use a blend of stabilizing materials. Frandsen and Arbuckle[57] listed some common stabilizer combinations to include CMC-carrageenan, CMC-gelatin, CMC-gelatin-carrageenan, and CMC-carrageenan-locust bean gum.

Today, modern stabilizer systems for ice cream contain predominately gum blends rather than a single hydrocolloid. Although most gums will produce some kind of desirable result in ice cream, combinations of gums are utilized to take advantage of specific, unique properties of certain individual stabilizers. No one gum possesses the "complete package" of functional tools that would allow its sole use in ice cream as the "perfect stabilizer". This finding explains why so many gums have been trialed in the ice cream "test vehicle" — to seek perfection is a basic human trait.

In view of the above, CMC is no exception. Although CMC imparts a variety of stabilizing functions to ice cream for reasons of cost, balance, synergistic interaction, whey-off or serum separation, and viscosity modification or control, CMC in combination with other colloids yields ultimately superior stabilization of ice cream.

Blihovde[71] recommended the combination of 1 to 12 parts of CMC with 1 part of carrageenan as an effective stabilizer system. Keeney[72] suggested a blend of CMC and carrageenan at 0.35% for a typical ice cream mix containing 10.2% milk fat, 12% serum solids, 13% sucrose and 0.15% mono- or diglyceride-type emulsifier. These gum combinations were not established out of capricious whim.

CMC is classified among ice cream additivies as a primary stabilizer. That is, it is the main functional choice for mix viscosity, body, heat shock control, proper meltdown and smooth texture. However, like other primary stabilizers, including guar and locust bean gums, CMC has the one drawback of causing whey-off or serum separation in the ice cream mix prior to freezing. Such separation readily occurs when the mix is aged or if the mix is shipped in flexible packaging or refrigerated trucks to be processed at another location. Needless to say, the whey-off phenomenon causes a lack of mix uniformity which drastically influences finished product quality and texture. To remedy this situation, a secondary stabilizer, carrageenan, is utilized. The carrageenan forms a weak gel which immobilizes the serum separation caused by the primary stabilizer. Thus a type of symbiosis exists. Use of a secondary stabilizer, carrageenan, is dependent on an adverse property of the primary stabilizer, CMC. However, the secondary stabilizer cannot functionally and economically replace the multifaceted role of the primary stabilizer. Both are needed.

Some manufacturers choose medium-chain-length CMC types instead of high-viscosity CMC grades to minimize whey-off. The greater preponderance of unsubstituted regions that characterize high-viscosity CMC types are believed to play a role in intensifying the whey-off phenomenon.

Ice cream mixes that are readily processed into the frozen state, such as in continuous plant operations, generally do not require a secondary stabilizer for whey-off prevention. Here, primary stabilizers alone, such as CMC, are perfectly adequate.

Stabilizers composed of various gum types serve other functions besides whey-off prevention. One of the classical problems with ice creams requiring more than usual stabilization is a deleterious mouthfeel, i.e., gumminess. However, without gums, far greater problems would arise. Finney[73] has devised a stabilizer system composed of microcrystalline cellulose (Avicel®), CMC, and various galactomannans that effectively stabilizes ice cream without giving an unacceptable mouthfeel. Using a balanced combination of gums, the finished product is reportedly spoonable at $-20°C$.

Ice cream stabilization has evolved into a highly sophisticated business which is handled primarily by stabilizer houses. Here CMC plays a major role. These firms compose and optimize gum/emulsifier blends along with other components to produce effective stabilizer systems and, as such, are consumers of CMC. These systems are then offered to independent dairies and processors in a ready-made and thoroughly tested form for addition to the ice cream mix. A typical system would consist of CMC-7HCF as the primary stabilizer along with guar or locust bean gum, carrageenan as the secondary stabilizer, 80/20 emulsifier, dextrose, salts, and anticaking agents.

The type of manufacturing process for ice cream influences stabilizer usage. Batch pasteurization still exists, but by and large, the more efficient high-temperature short-time (HTST) continuous process is preferred and utilized, especially for large-scale operations. As the ice cream industry advanced to more modern processes, stabilizers that performed adequately in batch systems were less effective in the HTST system. It seems that the shorter holding time characteristic of HTST reduced hydrocolloid functionality, especially with those gums not readily cold soluble. Typically, with HTST, the mix is heated to 175°F for 20 to 30 sec, whereas a batch process involves heating at 160°F for 30 min. Moss[55] first recognized the need for stabilizers having special properties for HTST use. Generally, 25% more stabilizer was required by HTST to maintain status quo with batch processes.[54,55] With the introduction of CMC, especially because of its excellent cold solubility, a particularly effective stabilizer was found which satisfied the requirements for HTST processing.[54,55] The gum could be added as a presolution or in the dry form, but far greater efficiency was achieved using the presolution technique.[50]

Dispersion has always been a problem with the use of hydrocolloids in ice cream mixes. The gum must dissolve under cold conditions and compete with other ingredients in the mix for available water. CMC is particularly effective in this capacity, and steps have been taken to even further improve its functionality. Landers[74] improved the dispersibility of CMC by coating it with mono- and diglycerides and found this particularly effective for ice cream mixes. Similarly, Steinitz[75] increased the dispersion speed of ice cream stabilizers by mixing them with propylene glycol and glyceryl monostearate.

The modern approach to improved dispersibility for CMC in ice cream mixes is the avalability of a coarse particle size from various manufacturers. This special granulation allows addition of the gum to cold viscous ice cream formulations under conditions of poor agitation without development of gum lumps which reduce stabilizer effectiveness and hamper processing. In situations where processors add bags of stabilizer blends directly to slow-moving vats of ice cream mix, use of coarse grind CMC is particularly advantageous.

2. Soft Frozen Dairy Foods

Soft frozen dairy products are characterized by a lower milkfat content (3 to 6%) compared to conventional hard ice cream. Included in this category of products are ice milk, soft serve ice cream and machine milkshakes. The latter two items have become tremendously popular in recent years, especially in fast-food restaurant chains.

From a texture standpoint, soft serve ice creams are self-defined. Milkshakes can best be described as "drinkable ice cream". They are enjoyed by older and younger generations alike. Milkshakes had their early beginning in the small-scale dairy bar/soda fountain shops frequented by "after-school clientele." Today, the milkshake has become highly commercialized in the U.S. marketplace where it may be encountered as an instant, canned, or machine-processed product, the latter, of course, being the dominant form.

Soft frozen dairy products employ CMC as a primary stabilizer for many of the same functions already discussed in hard ice cream. Because of the lower milkfat content, soft frozen products require somewhat higher usages of CMC to achieve desired textural characteristics. Inadequate stabilization tends to yield products that are coarse, weak, sandy, or icy. Arbuckle[50] suggested that stabilizers for soft frozen products be used in amounts ranging from 0.2 to 0.4%, whereas hard ice cream generally requires amounts just below this usage.

As soft serve and milkshake mixes are shipped predominately in liquid form to freezing sites (retailers), serum separation from primary stabilizers is a common problem and must be prevented with a secondary stabilizer. CMC-carrageenan combinations are routinely and effectively utilized in this capacity.

3. Instant Frozen Desserts

Do-it-yourself, instant frozen dessert mixes have been introduced to the marketplace on several occasions; however, these products have not gained the widespread popularity enjoyed by conventional, processed hard ice cream.

CMC is a particularly favorite stabilizing colloid for this type of dessert, because of its rapid solubility in cold water or milk as well as the many functional advantages of the gum already proven in regular ice cream.

A major problem with instant frozen desserts is formation of large ice crystals in the matrix during static freezing conditions encountered in home freezers. Commercial ice cream minimizes ice crystal formation by maintaining agitation and aeration during freezing. CMC's ability to prevent ice crystal growth in frozen systems is put to specific use in instant frozen desserts.

Baugher[76] developed a two-component instant frozen dessert comprised of an emulsion system and a dry mix packet which are combined with milk in a home mixer at high speed for aeration, and subsequently the blend is frozen. The dry mix contains a five-component stabilizer combination, including CMC type 9M31F, which functions to inhibit ice crystal formation, prevents mix stratification during freezing, facilatates aeration and interacts synergistically with each other and protein in the system to provide body and texture. Bundus[77] proposed a single package, dry, freezer dessert mix preparable with water which suggested the use of various cold-soluble hydrocolloids, including CMC, for prevention of undesirable ice crystal formation. Researchers at North Carolina State University developed a dry formulation for a frozen peanut dessert (imitation ice cream) which included 0.40% sodium carboxymethylcellulose for stability and texture control.[78]

4. Ices and Sherbets

Arbuckle[50] distinguishes sherbets and ices from ice cream by the following characteristics: (1) a higher fruit acid content, minimum of 0.35% which produces a tart flavor; (2) a much lower overrun, usually 25 to 45%; (3) a higher sugar content, between 25 and 35% which gives a lower melting point; (4) a coarser texture; (5) a greater cooling characteristic while being consumed, due to their coarser texture and lower melting point; and (6) an apparent lack of richness due to lower milk solids content. Sherbets are further differentiated from ices in that the latter are devoid of milk solids.

Stabilizer requirements for sherbets and ices are significantly greater than other frozen desserts because of acidity, lower total solids, lower overrun, and greater dependency on cooler temperatures for maintenance of uniformity. In choosing a stabilizer for these frozen products, concern should be given to the gum's effect on overrun, syrup drainage, and body (crumbliness).[57] Stability of the gum in an acidic medium, expecially when extra acidulant is added to the product to improve tartness, is an additional requirement.

Effective stabilization of sherbets and ices is far more crucial than for ice cream because of great danger of sugar separation or migration of other components with minor changes in product temperature. During freezing of products, water tends to crystallize out first as large ice crystals, and solutes such as sugar, color, and flavor concentrate into an undesirable sticky, gummy residue. "Bleeding" is a defect in sherbets whereby sugar syrup settles to the bottom of the container. This is attributed to insufficient amounts of stabilizer. Conversely, products such as ice pops tend to undergo color and flavor migration when outer product surfaces melt from heat shock or exposure to warmer temperatures in the mouth. Thus a hydrophilic stabilizer must be used in sherbets and ices to improve texture by preventing ice crystal growth and to maintain uniform distribution of flavor, color and soluble solids during freezing, storage, and eating.[79]

Selection of the proper type and amount of hydrocolloid is essential to the stability of ices and sherbet. Burt[66,81] found that 0.1 to 0.2% low viscosity CMC effectively stabilized ice products.

Surface encrustation from solute migration and excessive coarseness in ices and sherbets are caused by insufficient stabilizer. A sticky body is a consequence of too much stabilizer or too much sugar.

Best stabilization of ices and sherbet is obtained when a combination of gums function to cause a partial gelling at cold room temperatures.[50] Braverman[85] showed that combinations of either low-methoxyl pectin and CMC or carrageenan and CMC in freeze-it-yourself pops provides the composition with homogeneity, stability, and consistency in both frozen and unfrozen states. These gum combinations allow solid particles, such as cocoa, fruit, titanium dioxide, and artificial color, to be uniformly suspended in the nonfrozen state and, when frozen, the product is chewy and virtually non-icy. The gum systems impart to the admixture a thickness or texture (at room temperature) which is on the verge of being semisolid, i.e., a pudding-like or gel-like consistency.

Citric acid is the most common acidulant for ices and sherbet. It is the acid component that differentiates ices and sherbets from ice cream. Hydrocolloids employed in acidic frozen products must withstand a lower pH environment. High DS or uniformly substituted types of CMC offer optimum stability and effectiveness under these conditions. For example, CMC-7H0F or CMC-7H3SF are recommended by one manufacturer for ices and sherbets.[86]

Slush-type products may be considered as the result of cross-breeding between a frozen dessert and a beverage. Typically these products are dispensed from a machine and range in consistency from a viscous semifrozen drink to a sherbet-like texture. Compositionally, slush products contain ingredients similar to sherbet and ices; however, the total solids content is much lower.

The major problem with slush products is undesirable development of large, shale-like ice crystals during operation of the slush machine with concurrent settling of soluble solids into a syrupy layer. Stabilizers, such as CMC, are utilized to prevent these problems.

Marulich[87] found that pectin in combination with CMC is effective in regulating ice crystal structure of slush to a desirable sherbet-like consistency. Homler et al.[88] describe a process where ionic cellulose gums help prevent separation in slush beverages with a gaseous overrun. A combination of 0.45 parts CMC-7LF and 0.1 part CMC-7H4F at 0.64% was utilized in an imitation orange slush drink. Rubenstein[89] used 0.16% guar gum and 0.1% CMC in an ice water mix which was combined with a liquid flavor base containing whipping agent (soy albumin) to produce a slush product with the richness of a milk drink.[89] Le Van et al.[90] utilized CMC in the preparation of a homogeneous frozen concentrate which upon dilution with water, carbonated water or milk yielded an acceptable slush beverage.

High viscosity CMC types are more effective as cryoscopic modifiers in slush products; however, limitations exist in that too much CMC tends to impart a gummy mouthfeel and inhibits dispensing of product from slush units.[91]

5. Variegated Sauces and Ribbonettes for Ice Cream

CMC serves to stabilize and thicken variegated sauces and ribbonettes which are specialized flavor preparations for ice cream. These products are injected into the ice cream as they evolve from the freezing apparatus. The CMC functions to thicken, prevent sugar migration or crystallization, and provides flow control. CMC-7H3SF is mainly recommended by one manufacturer for these specialized ice cream flavors.[86]

6. Milk Beverages

CMC serves as an effective stabilizer in both acidified and neutral pH milk beverages. Important functions exhibited by CMC in these applications include a thickening effect, which in turn imparts desirable body, mouthfeel, and texture; and interaction with milk protein (primarily casein), which provided synergistic viscosity development, a protective colloid effect on the milk protein at acidic pH, and heat stability of the milk protein under conditions that would normally promote denaturation.[46,97] The effect upon the system is dependent on the composition and concentration of the milk proteins, the concentration and type of CMC, pH, and temperature.

Prepared milk drinks such as eggnog, milk shakes, infant formulas, and to some extent, chocolate milk are dairy beverages that utilize the stabilizing and texturizing properties of CMC.

Eggnog is a seasonal drink, popular during Christmas time, that is composed of milk, liquid egg yolks, sweeteners, rum ether, and various spices, such as nutmeg. This product customarily employs high-viscosity, uniformly substituted CMC for acceptable smooth mouthfeel and viscosity. Manufacture of eggnog by HTST pasteurization is now a standard method in the dairy industry. As such, the basic recommendations previously given for ice cream stabilizers usually hold true. When high-viscosity CMC is used in eggnog, it is preferable to include a secondary stabilizer such as carrageenan to prevent mix separation.[54]

Ready-to-serve, sterilized milkshake drinks in cans or flexible aseptic packages require effective stabilizers similar to those utilized in machine-processed milkshakes. CMC, because of its thickening action and favorable reactivity with milk protein (especially when the complex is heated), is a popular choice for these products. Smith[94] developed a packaged liquid milkshake mix containing a viscosity control agent comprised of one part carrageenan and from 2 to 5 parts of cellulose gum. After addition of the viscosity control agent, the product is then homogenized, cooled, filled into sanitary cans, and sterilized by a conventional continuous HTST method; 265°F for 120 sec, then cooled to 80°F. Products prepared by this technique have extended shelf life at ambient temperature. After refrigeration and agitation, the can-dispensed shake has the texture and mouthfeel typical of a soda fountain shake.

Stewart et al.[95] engineered a process whereby CMC is injected as a presterilized aqueous solution into a milkshake mix previously processed by HTST. It is claimed that this technique minimizes sedimentation from protein coagulation in the finished product to a level far below that encountered in other processes. The heat stability of the CMC-casein complex is involved in this process.

Traditionally, chocolate milk is stabilized with carrageenan at very low usages (0.02 to 0.035%) which forms a weak gel that suspends the cocoa particles. Suspension of cocoa is the primordial function and oldest known application of carrageenan for milk drinks. However, further addition of carrageenan for viscosity purposes is uneconomical and quite detrimental since gel formation is so intense and brittle that a separation occurs. Therefore, other gums such as CMC, guar, or starch supplement the carrageenan as bodying or viscosifying agents.

In other processed milk products, such as condensed whole milk or skim milk, CMC has been reported to prevent the formation of large, undesirable lactose crystals.[96] This is accomplished by preparing a seed material whereby CMC and crystalline lactose are intimately ground together and incorporated into a small amount of concentrated milk product, such as spray-dried milk powder. Crystals of about 1 μm in diameter evolve from this technique. When added during the cooling cycle in the production of condensed milk, the admixture creates a ''tempering effect'' in the final product whereby 50% of the lactose crystals are 0 to 5 μm in diameter and 50% are above 5 μm.

CMC facilitates the preparation of acid milk beverages by forming a stable complex with casein in the pH range of about 3.0 to 5.5.[92,97] Figure 22 illustrates the basic interaction. Normally, acidification of milk to the pH range of 3.0 to 5.5 causes the casein to precipitate, i.e., curdling. Heating of acidified milk would intensify curdling. Since the CMC-casein complex is also heat stable, milk may be acidified, such as fruit-flavored milk, and subsequently heated to produce a pasteurized or sterile product with extended shelf life.

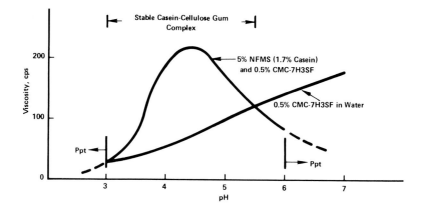

FIGURE 22.   Viscosity effect of CMC-casein complex at varying pH.

Asano[41] investigated the stabilizing mechanism of fruit-flavored milk by studying the interaction between CMC and various milk proteins with electrophoretic analysis, optical density determination, and viscosity measurement at different pH values and temperatures. He observed that maximum viscosity for the interaction of CMC and milk proteins occurs between pH 4.0 to 5.0 which is the optimum pH region for desirable organoleptic appeal of fruit-flavored milks. The driving force of the interaction between CMC and milk proteins was attributed to electrostatic attraction between opposite-charged groups on the protein and polysaccharide chains.

Shenkenberg et al.[27] describe the use of CMC to prepare a nutritious, milk-fruit beverage, containing about 2 parts milk to 1 part orange juice. A dry blend of sucrose and CMC is first added to casein containing milk. The mixture is aged for not less than 10 min below 90°F to allow complex formation between the casein and CMC. Fruit juice may now be added to the complexed mixture after which the system is pasteurized, homogenized, cooled, and packaged. Low, desired viscosity in the product was obtained by using approximately 0.2% of a medium viscosity 0.7 DS CMC.

A novel, uncoagulated milk-apple juice drink stabilized with CMC was prepared by Nishiyama.[143] Essentially, a CMC presolution is prepared at 75 to 80°C, apple juice is added and the mixture is acidified preferably with citric or lactic acid. This "apple juice composition" is combined with sugar syrup and milk (in critical ratios), pasteurized, homogenized, and filled. The technique of adding a composite of CMC-apple juice-acidulant to milk distinguishes this invention from others, and it is claimed that enhanced stability results from the methodology.

Campbell and Ford[144] mathematically describe the utilization of CMC in the preparation of a stable acid milk beverage. Accordingly, this invention provides for a pasteurized mixture in water comprising CMC, milk solids, and an edible acid whereby enough CMC is employed to satisfy the relationship:

$$Y = 0.4x + 0.2$$

Where Y is the percent by weight of CMC salt and x is the percent by weight of milk protein. The edible acid must bring the pH of the beverage to a desired 2.5 to 4.0. Deviation of ingredient levels away from those prescribed in the equation produces a less desirable product. A low-viscosity-grade CMC (100 cps as a 3% w/w solution) in a DS range of 0.7 to 0.8 was specified for this particular beverage.

The advent of high-fructose corn syrup has had a definite impact upon many segments of the food industry, including dairy products. Daehler[142] describes an acidified milk drink which utilized high-fructose corn syrup for sweetness and several types of gums to stabilize the beverage below the isoelectric point of milk protein. A combination of calcium carrageenan and acid-resistant CMC produced the best results.

Because the CMC-milk protein interaction focuses on a protective colloid effect with maximum viscosity development at acidic pH, most manufacturers prefer medium-viscosity types of CMC to minimize consistency since high-viscosity types of CMC tend to overpower the system.[86]

Morgan et al.[98] commented on the use of CMC in conjunction with guar gum and carrageenan to prevent viscosity loss during heating of commercially sterile fermented milk products. Cultured dairy products experience viscosity breakdown during refrigeration because of ongoing microbial action. Elimination of active microorganisms by heat sterilization without proper stabilizers tends to cause thinning down, loss of body, and watery separation. In this invention, these workers admixed various gums, including CMC, in water which was then added to the fermented products prior to heat sterilization to avoid the adverse effect described above.

Cajigas[99] developed an instant yogurt drink which utilized the stabilizing ability of CMC upon milk protein at acidic pH. This product consisted of deactivated yogurt powder, a dried *Lactobacillus* culture which is activated upon reconstitution and simulates natural yogurt, an acidulent, and CMC as a preferred thickener and stabilizer. Of particular interest, the inventor cites an optimum particle size for hydrocolloids as a key factor in making an acceptable finished drink.

Nutricia[100] also described an acid milk beverage prepared by mixing whole milk, skim milk, or cream with a lactic acid bacteria culture, at least 10% fruit juice, and sufficient CMC to keep the casein dispersed in finely divided form. The mixture may be pasteurized or sterilized.

Although the CMC-milk protein interaction at acidic pH is of major importance to foods, the complex will in fact form at other pH values with different utility. Above a pH of approximately 6, CMC forms a complex with milk protein that may be removed as a precipitate.[38,93]

Maximum precipitation of protein takes place between pH 7 and 8, and high-viscosity CMC is preferred in this case.[92] The reaction is calcium dependent, and sequestering agents will block the precipitation.[93] This interaction has been utilized for the commercial separation of casein from milk.

Complex formation between CMC and milk protein with subsequent precipitation also occurs at more acidic pHs, approximately 3.5 or below. Hill and Zadow[35] showed that the extent of protein precipitation from cheddar cheese whey as insoluble complexes with CMC depended on DS, pH, and the ionic strength of the system. Abdel Baky et al.[101] demonstrated the recovery of protein from unsalted and salted whey fractions of Domiati or Ras cheeses by complexing with CMC. The recovered whey protein/CMC complex was then utilized to fortify fresh cheese and cheese milk to increase yield, moisture, and protein content. Equal volumes of whey and 0.025% CMC solution were mixed to achieve complex formation.

Both groups of workers discussed above cited an optimum pH requirement of 3.2 to 3.4 for onset of precipitation of the whey/CMC complex.[35,101]

B. Baked Products

Compositionally, baked products are constructed of fundamental food components such as protein, fats, sweeteners, fruits, flavors, cocoa, salts, grains, yeast, and water. By application of heat to an integrated system of these components, such as a dough or batter, a metamorphosis occurs which changes the system into a finished bakery product characterized by new and different properties and markedly improved eating qualities.

Many complex interactions occur during the "baking transformation." Among these, the effect of heat on moisture content is of primary importance. Moisture, whether it is lost or retained in a free or bound state during or after baking, has profound influence on the texture, flavor, appearance, and shelf life of baked products. Therefore, control of moisture in baked products is of utmost importance.

CMC, because of its excellent ability to serve as a moisture regulating agent, is a key ingredient in a wide variety of bakery products. The gum is employed in cakes and other chemically leavened mixes, icings, fillings, doughs, meringues, glazes, and other specialty bakery items. Besides moisture control, improvement in body, repression of sugar or ice crystal growth, film formation (surface glaze), texture modification, interaction with starch and proteins such as gluten, volume development, batter viscosity control, and structural enhancement are other important functions of CMC in bakery products.

1. Cakes and Other Chemically Leavened Products

Packaged ready-to-use cake mixes represent the ultimate in convenience, reliability, and quality for the home baker compared to the laborious chore of putting together a mix from scratch. The modern boxed cake mix has successfully evolved to meet the time-pressured schedules of a fast-paced society and, as such, occupies a major niche in supermarkets today. Vast improvements in packaged cake mixes allows the homemaker to turn out a product equivalent to commercial cakes simply and easily. As such, this situation represents increased competition for the commercial cake baker who formerly enjoyed a big advantage over consumers in that he could control his ingredients much better than the homemaker. This advantage has diminished greatly.

Highly specialized cake ingredients, like CMC, are largely responsible for the convenience and quality aspects of packaged cake mixes. Pyler commented that "both batter character-istics and final cake quality can often be improved by the use of hydrophilic materials, such as some chemically modified or pregelatinized food starches and certain vegetable gums."[102]

The foremost function of CMC in cake mixes is that of a moisture imbiber. Nothing is more objectionable, organoleptically, than a dried-out cake. Rapid onset of this problem reduced shelf life which proves uneconomical for manufacturers as well as consumers. CMC, because of its excellent affinity for moisture, is very effective in preventing the most for-midable of cake defects — dryness. High DS or high molecular weight CMC types are preferred choices for cake mixes, since these CMC species exhibit maximum moisture binding.[86]

Other performance properties imparted by CMC in cake mixes include improvement of volume, symmetry, grain, and texture (crumb softness); facilitation of single stage batter preparation; provision of tolerance by helping to protect the mix against leavening loss, inaccurate liquid addition, and heat damage; allowance of increased sugar levels as found in modern high-ratio cake systems (excess sugar provides overtenderization which is com-pensated for by the toughening action of CMC); increased and controlled batter viscosity and uniformity; and regulation of batter specific gravity. The ability of CMC to influence this last property is very important as batter specific gravity serves as a "barometer" for correct finished cake properties and eating qualities. Non-optimum batter specific gravities tend to yield cakes with lower volumes, denser grain, surface dips, friable structure, and tougher eating qualities.[102]

Many workers have reported on and established the multifunctional role of CMC in cake mixes. Jaeger[103] first recognized the need to stabilize the moisture in a dough or batter before, during, and after baking so that improvement in palatability and retention of freshness results. This worker suggested "emulsifying agents", of the mucilaginous type, as an effective means of achieving many desired characteristics, especially moistness and fresh-ness, in finished baked foods. Unfortunately, CMC was not used in this study, probably because the gum was in its infancy as a food ingredient at that time.

Bayfield[104] demonstrated improved properties and higher cake scores with addition of small amounts of CMC to high ratio white layer cakes. The shelf life of the CMC cakes was longer than those baked without CMC. Young and Bayfield[105] tested various hydrophilic colloids in white layer cake. Among the materials tested, the use of 0.1% CMC-7HP (a pharmaceutical grade) produced the highest scores during various storage periods. Cakes were stored unwrapped to accelerate the staling process and were scored after 16 and 48 hr of storage. Dubois[106] cited the role of CMC in batter cakes as a viscosity stabilizer and moisture retainer. Recommendation was made to use between 0.25 to 0.375% gum based on flour in the batter with a concurrent increase in water at a rate of 30 to 50 oz for each ounce of cellulose gum added.[106]

The work of Elsesser[107] pioneered the use of CMC in convenience cake mixes designed for the homemaker. The gist of this invention specifies incorporation of CMC with non-shortening components (flour and sugar) prior to combining with an emulsified shortening. In this way, greatly improved ease of batter preparation occurs such that minimum physical force is required for rapid viscosity development and better batter uniformity. The system allows a single stage liquid addition technique which is far simpler than time-consuming, energy-intensive creaming methods or multistage liquid additions. Use of CMC in this single stage mix allows ample air incorporation as evidenced by low batter densities, improved mix tolerance to overbeating and a reduction in shortening (emulsified) which permits a higher moisture level in the batter. Consequently, quality, high volume cakes with freshness and excellent eating properties may be easily prepared. Another beneficial function of CMC in this dry mix is that the gum does not introduce premature rapid thinning of the batter during baking which gives uniform leavening, avoids moisture gradients, and offers opportunity for suspension of "inert" materials like flavoring chips or chunks. Preferred usages for CMC in this disclosure are 0.06% for white or vanilla cakes, 0.4% for devil's food cake, 0.2% for yellow cakes, and 0.3% for spice cakes.

Hager and Lowry[108] suggested the inclusion of CMC into the shortening phase of a single stage cake mix as a more effective dispersion technique for the gum. By subjection to a crushing and shearing operation, the resultant mix is characterized by a major part of the sugar crystals thereof being fragmented and mechanically bonded to the flour particles. This composite is simultaneously coated by a thin film of shortening containing homogeneously dispersed CMC. Cakes utilizing this ingredient technology exhibited excellent volume development and were superior in symmetry, grain, and organoleptic qualities compared to earlier cake mixes requiring more difficult reconstitution.

CMC is beneficial in the prevention of other types of cake defects. Trimbo et al.[109] described a "ring" migration during baking which led to rough surfaces, cratering and weak structure. The phenomenon was attributed to natural convection currents which are caused by differences in batter density due to lateral and vertical temperature gradients within the heated system. Prevention of ring formation was accomplished by addition of 0.5% CMC (based on flour weight) which bound the water, increased batter viscosity, and inhibited lateral batter flow.

Young and Bayfield[106] observed that 0.6% CMC-7HP overpowered white layer cakes and yielded a detrimental gummy texture. Too often, excess CMC in a food system promotes adverse textures and this problem is still encountered today. However, Miller et al.[110] showed that, without CMC, cakes develop a gummy layer which has greater density, higher moisture content, and a higher degree of starch gelatinization than normal crumb cake. This gummy layer is a result of moisture movement to localized cake sections from other areas of the batter. These workers demonstrated the ability of CMC (0.5%) to prevent moisture transfer and subsequent development of gummy layers in cake batters heated in sealed cans.

Other varieties of cakes and cake-related products rely on CMC for stabilization. Goodman et al.[111] cited the use of CMC for stabilization and texture enhancement in a fluid matrix which is admix to a second fraction of prebaked farinaceous material then freeze-hardened to produce a frozen cake product. Gass[112] developed a dry cake topping comprised of sugar, shortening, and a water-binding agent (pregelatinized starch, CMC, and alginate) which is placed on top of the batter before baking. During baking, the topping ingredients rise with and remain on top of the batter. The water-binding agent absorbs water during baking such that a glaze-type topping results on the finished product. Weigle[113] described the addition of 3.0 to 6.0% high-viscosity CMC in strawberry flavor bits to help them "suspend" in a cake batter. Tests indicated that at least 3% CMC is necessary to prevent settling out of the bits and subsequent uneven flavor distribution in the finished cake.

A brownie mix is another type of chemically leavened, convenience bakery product, high in cocoa, which may best be described as an anticake. Brownies, especially the fudgy type, tend to have low volume development, tight grain and a chewy, gummy texture; are dense rather than light; are sticky; develop extremely viscous batters; and are high in fat, solids, and moisture. These properties would not be desirable in regular cakes. Small amounts of CMC have been successfully tried out in brownie mixes to actually enhance the gummy chewy texture and to extend shelf life, as these products readily dry out.

In donut mixes, chemical or yeast leavened, it is claimed that CMC improves yield, deters fat absorption during frying, enhances structure and crumb tenderness, binds moisture, and promotes longer shelf life. A usage of 0.25% CMC in a donut mix has been suggested to impart these desired properties.[10]

The utilization of CMC in cookie mixes, coffee cakes, pound cakes, pastries and other sweet goods, institutional cakes, muffin mixes, general baking mixes, and continuous process industrial cakes has not been extensively explored. These products offer great potential for application of CMC whereby functional gum properties would enhance organoleptic qualities and extend shelf life. Commercial products shipped interstate or internationally have limited road and shelf life. Hence, any means of increasing this shelf life would yield substantial monetary savings to manufacturers.

2. Icings, Frosting, and Glazes

Just as no gift would be complete without the appropriate wrapping and ribbon, icings, frostings, and glazes provide the necessary "decorative packaging" for bakery foods. The importance of these bakery adjuncts cannot be overemphasized. Dubois[114] classified coated bakery foods as "impulse" items. It is the icing or glaze that is the first point of contact between the consumer and bakery product and thus influences the decision to buy. Apart from the eye appeal requirements, these embellishments must contribute to the overall eating qualities of bakery items as well.

Icings, frostings, and glazes consist predominately of sugar and water with lesser proportions of fat, milk solids, egg solids, stabilizers, flavors, salt, and colors. Fat content helps distinguish between product types. For example, "flat" icings or glazes for sweet rolls or donuts contain 0 to 1% fat, cupcake icings are 2.5 to 5.0% fat, and cake icings may possess 10% or more fat.

Physically, an icing or glaze is a two-phase system consisting of small sugar crystals dispersed in a sugar syrup (saturated sugar solution).[114-116] The quality of the icing or glaze is therefore determined by the degree of control over this sugar-water system during product preparation and storage. Lipman stated that "hydrocolloids are used to control the delicate balance between the dissolved sugar and the suspended sugar, modifying the crystallizing characteristics and solubility of the sugar in aqueous medium, and thereby stabilizing the size of the sugar crystals.[117] Birnbaum[116] described the chief role of gum additives in icings as water immobilizers. Although water is necessary for the spreadability of the icing, it also causes icing quality deterioration when uncontrolled.

CMC, because of its excellent ability to bind moisture and retard undesirable sugar crystal growth, is functionally well suited and widely used to stabilize bakery coatings. CMC aids in the handling, spreading, and pliability of these products, prevents cracking, maintains gloss and true color, imparts a smooth velvety character, and most important, it inhibits gritty texture development and drying out of the coating. When dry, hard, and brittle, icings lose adherence to bakery products and fall off easily.

High-DS, uniformly substituted, nonthixotropic CMC types are preferred for bakery coatings for maximum moisture binding and smooth texture.[86]

Another advantage of CMC in icings is that icings do not separate when stored as long as 24 hr and they can be rewhipped for reuse.[10] CMC helps retard transfer of moisture from the atmosphere to the icing, from the icing to the baked item, or from the baked item to the icing, thus reducing stickiness.[115]

Various developers have utilized CMC as a stabilizer in icing formulations. Wagner[118] prepared a stable, boiled icing-base dry mix with a blend of egg albumin, calcium sulfate, sodium aluminum sulfate, powdered sugar, starch, flavor, and CMC. Butler[119] used CMC in an instant frosting mix which included a whipping agent for aeration. Ganz[120] described an icing formulation stabilized with CMC and vital wheat gluten in combination. The protein is water insoluble, but hydrates when admixed with CMC and incorporated into the icing composition. This stabilizer combination prevented "bleeding" (liquid separation), drying, hardening, cracking, sugar crystallization, and minimized adherence of the icing to cellophane wrappers. Grossi[121] claimed that the use of CMC in a dry chocolate icing mix gave improved color stability.

Several workers list agar as the most basic gum for nearly all icing stabilizer systems and cite its excellent gelling property as being "unsurpassed" for immobilization of water in icings.[114-116,122] Similar to ice cream stabilizer systems, many icing stabilizer combinations are composed of a blend of gums (primarily agar), emulsifiers, fats, whitening agents, and flavors. These systems and their components have been described in detail by Birnbaum,[114] Wheeler and Endres,[116] Lipman,[117] and Dubois.[123] However, utilization of agar alone is not a panacea for icing stability problems. Morley[124] stressed the need for multigum stabilizer systems for icings and glazes to achieve trouble-free performance.

Specifically, quality defects in icings stabilized with agar as the sole hydrocolloid arise in that atmospheric moisture condenses on the gel, causing sugar dissolution (stickiness), or in low humidity conditions, moisture will easily evaporate from the agar gel which causes hardening and loss of pliability. A supplementary gum, CMC, was recommended to alleviate these problems.

Dubois[114] describes the so-called single strength stabilizer (10% usage) or double strength stabilizer (4 to 6% usage) for icings and glazes which customarily contain agar and smaller amounts of other gums for well rounded stability.

Glazes for honeybuns and doughnuts employ CMC at about 0.1 to 0.2% to give a hard, glossy, attractive finish to products of this type.[10]

Of recent vintage, ready-to-spread, processed creamy frostings in plastic cannisters have gained tremendous popularity in the consumer marketplace. These convenience products complement the box cake mixes described previously.

Ready-to-spread frostings are rather complex emulsions, high in sugar, water, and shortening, which require a portfolio of gums, emulsifiers, and preservatives to maintain integrity, texture, and microbial stability with varying storage temperatures. Generally, these frostings are stable at ambient temperatures when sealed, but refrigeration is recommended after opening of the cannister.

Cellulose gum is routinely used in ready-to-spread frostings, usually with pectin or agar, to help bind moisture, control sugar crystal growth, stabilize the emulsion, aid pliability, and to help maintain the spreadable consistency of the product.

3. Meringues
Aerated meringue toppings are popular adornments for baked goods, especially pies.

Lemon meringue pie and key lime pie, the latter being indigenous to the Florida Keys, are classic bakery examples that utilize meringues.

CMC is commonly added to meringue products to stabilize the foam, enhance volume development, provide tenderization, impart a smooth, shiny appearance, maintain uniform structure, allow ease of cutting, and prevent weeping. This latter problem is the most common defect in meringues and high-viscosity, high-DS CMC types are preferred for maximum moisture binding.[86]

A typical hot-process meringue stabilizer consists of 63% dextrose, 15% cornstarch, 15% high-viscosity CMC, and 7% agar. For a cold-process meringue, the agar must be eliminated and a formulation can be prepared using 69% dextrose, 18% cornstarch, and 13% high viscosity CMC.

A dry mix meringue powder recipe containing CMC for stabilization has been suggested by Hercules, Inc. for application to regular meringue, cold marshmallow meringue, and boiled marshmallow meringue.[125]

4. Pie Fillings

Fruit-based and cream style (soft) pie fillings traditionally utilize starch to give the desired gel structure, consistency, and cost-effectiveness. However, some drawbacks exist in that copious usages of starch in these fillings mask flavor, impart a slimy, starchy mouthfeel, give a dull appearance, allow shrinkage and cracking, and weepage may occur. In pie fillings dominated by starch for thickening, CMC is used to modify and firm texture, minimize or prevent syneresis, add sparkle and sheen, and reduce starchy mouthfeel while maintaining desired consistency.

Starch-based lemon pie fillings often show cracking or pulling away from the sides of the pie crust during shelf storage. In lemon pie fillings containing about 4 to 8% starch, addition of 0.4 to 0.5% high-viscosity CMC "cements" the matrix to eliminate this defect.

In fruit pie fillings, CMC is used to thicken the fruit juices to the proper consistency for handling, heating, and prevention of fruit flotation or settling prior to placement in containers or pie shells.

Moyls et al.[126] found CMC to be a satisfactory stabilizer for canned berry and berry-apple pie fillings under accelerated storage conditions. Bisno[127] found low levels of CMC effective in preventing syneresis of pie fillings thickened with cornstarch alone.

Kunz and Robinson[128] investigated CMC in both canned and frozen peach and cherry pie fillings. The addition of 0.3 to 0.5% CMC alone or in combination with modified waxy maize starch resulted in peach pie fillings with a clearer, less milky liquid phase after 10 months of storage. In cherry pie fillings, CMC usage was maximized at 0.2% by these workers, otherwise a gummy texture was obtained. When pie fillings were frozen, incorporation of CMC decreased watery separation encountered after thawing.

Instant pie filling mixes reconstituted with cold milk utilize CMC in combination with starch for improved body and rapid texture development.

5. Bread and Other Dough-Derived Bakery Products

From Biblical times to the present day, staling of bread has been and continues to be the baker's nemesis. Everything possible must be done by the baker to delay the staling process and retain good eating properties in bread. In modern large-scale bread production, a high return of "stales" must be minimized since serious monetary losses will obviously be incurred.

Besides moisture, many other ingredients in bread contribute to the staling phenomenon, including gluten, pentosans, emulsifiers, and particularly, starch.

Conversely, many ingredients have been investigated by food scientists for antistaling functionality and evidence exists that hydrophilic colloids as additives will prolong the shelf life of white bread.

Schuurink[130] showed that addition of 1 to 2% CMC increased water retaining capacity and volume of doughs. Bayfield[131] confirmed and expanded the work of Schuurink by investigating the performance of various viscosity types of CMC in bread baked by a straight dough process. It was noted that addition of different types of CMC produced significant changes in the mixing behavior of the dough. Adjustment of water usage was necessary for optimum absorption and to give the dough the proper ''feel'' when CMC was included.

After 5 days of storage, bread with CMC possessed approximately 1.4% more moisture that the control and was definitely softer. An increase in bound moisture was believed to be at least partly responsible for the improved shelf life of bread containing CMC. It was also hypothesized that CMC in solution may serve to coat starch granules and thus delay the staling process which has been associated with starch components. Van Haften[132] suggests a more recent mechanism for staling prevention with emulsifiers whereby these ingredients complex with starch fractions as clathrates to inhibit starch crystallization or retrogradation known to occur during staling. It is speculated that CMC may play a similar role in retarding of staling, but more work on starch-CMC interactions will be required to substantiate such a mechanism.

Dubois[137] describes a bread additive consisting of 20 to 30 parts vital wheat gluten, 3 to 5 parts hydrophilic colloid (CMC), and 0.01 to 0.025 parts oxidizing agent which gives improved bread having increased loaf volume, desirable grain, texture, and keeping qualities. Unfortunately, bread is classified presently among standardized bakery products and as such CMC is precluded for use as an additive ingredient.

Unlike bread, where softness is desirable, baked foodstuffs such as pretzels, crackers, certain biscuits, and ice cream cones or wafers must retain crispness to be appreciated. Softening of these products from moisture absorption impairs their mechanical strength and may result in eventual disintegration. For example, ice cream wafers enclosing frozen dessert readily absorb moisture when subject to freeze/thaw conditions. Von Rymon Lipinski[133] disclosed a process for modification of CMC by crosslinking the polymer with heat or chemical agents to create an ingredient that prevents softening of baked flour containing products by absorbing incoming moisture. However, food approval of modified CMC, at least in the U.S., has not as yet been sanctioned.

Ready-to-bake frozen dough products, such as biscuits or doughnuts, may be subject to freeze/thaw stress which causes undesirable ice crystal growth (similar to ice cream) that influences finished product quality. It has been suggested that a small amount of CMC will effectively reduce the rate of ice crystal growth in frozen doughs.

Another reported use for CMC is in pizza dough mixes. Microwave cooking tends to drive moisture from the sauce into the crust, creating sogginess. The CMC acts as a ''sponge'' to bind the water and maintain a crisp crust.

A recent utilization of CMC is in masa-derived Mexican foods such as tortillas, tacos, and enchilidas. Masa dough is typically prepared by lime-steeping whole corn kernels overnight at 180°F in a process known as nixtamalization. After draining off excess fluid, residual corn is ground into a tortilla dough commonly referred to as masa having about 50 to 65% moisture content. Bulk dough is then rolled into flat sheets, cut to the desired tortilla shape, and baked at 750°F for about 45 sec to give a product having 40 to 50% moisture.

The tortillas are not sterile and are usually refrigerated or frozen. Typically, in the home, these products are mildly fried, meat filled, rolled, covered with sauce, and baked. To produce this end product, the tortilla must remain flexible for rolling and stay rolled during the preparation and baking process.

In an alternate procedure, the whole tortilla is fried in deep fat for about 30 sec in a U-shaped form to produce a hard, brittle taco shell of very low moisture content (2 to 4%). These taco shells are shelf stable at ambient temperature.

Soft tortillas are quite prone to staling or drying out and they must retain structure and flexibility during processing and consumption. Due to the much lower gluten content of corn flour compared to wheat flour, retention of moisture, structure, and flexibility in tortilla products is quite difficult. CMC is utilized in soft tortilla products to retain moisture, retard staling, maintain flexibility, and supplement structure formation due to the lack of sufficient gluten. Rubio[134] describes the incorporation of CMC into tortilla products to provide improved resistance to staling and moisture loss. Tims[136] cites a synergistic combination of vital wheat gluten and various hydrophilic gums in tortillas to improve flexibility, retain moisture, deter starch retrogradation with subsequent staling and to maintain structural integrity. The wheat gluten or gums alone were not as effective in providing the described functionality.

Low moisture, U-shaped taco shells are quite brittle and are readily subject to breakage in transit. Small amounts of CMC have been used in such products to provide some flexibility and help reduce breakage of the fragile shells.

6. Batters and Coatings

CMC has been employed in various batters and coatings for viscosity development and control, adhesion, suspension of solids and as an oil barrier during deep fat frying of coated products.

Complete, reduced-calorie pancake batter mixes requiring water only make use of CMC to add body and viscosity to the batter while allowing reduction of caloric contributing components such as starch, shortening, or dextrose. The gum also helps to decrease the "runniness" of the batter as it is placed on the griddle.

Breading batters are typically high in total solids ($\sim$ 40 to 50%) to maintain desired consistency. These solids are expensive. Rapid turnover and depletion of batter from reservoirs in coating operations adds to the cost factor. CMC provides an economic function to batters whereby solids may be partially reduced to save money while the gum builds back the necessary viscosity in the batter. Although not a true suspending aid with a yield value, the thickening effect of CMC in batters retards sedimentation of solids which leads to handling problems during processing and lack of uniformity when the batter is coated onto products.

In a related application, Suderman et al.[136] demonstrated that incorporation of CMC into a commercial breading mix gave significantly improved adhesion to milk-battered drumsticks compared to a control mix without gum. CMC was also significantly better than several other gums at the same usage amount for improving breading adhesion.

7. CMC Interaction with Bakery Ingredients

CMC undergoes some rather interesting interactions with various bakery ingredients for improved functional performance. Several of these interactions have not been extensively investigated, are not completely understood, and would make superb topics for dissertations within academia.

Egg and its components are key bakery ingredients. They are very functional in bakery systems, but are also expensive and perishable. For these reasons, replacement, extension, or improvement of eggs is an ongoing quest in bakery items and other food systems.

Chang[138] developed an egg albumen extender comprised of a whey protein concentrate and 0.5 to 15% sodium carboxymethylcellulose. The use of the carboxymethylcellulose in combination with the whey protein concentrate in baked goods such as cakes overcomes the deficiencies in cake texture and volume normally encountered in replacing egg albumen with whey protein alone. It is claimed that up to 75% of the albumen requirements in baking applications such as cakes, sweet doughs, biscuits, pancakes, doughnuts, and the like can be made with this invention.

Ganz dried egg white and CMC to yield a product having superior foaming, binding, and stabilizing properties in bakery products compared to lesser functionality obtained from simply dry mixing the two ingredients.[43]

Synergistic interaction between CMC and vital wheat gluten for stabilization of icing formulations has previously been described.[120]

The effect of CMC on starch gelatinization or retrogradation in bakery systems is relatively unexplored at this time and further work is needed to elucidate these interactions.

Roundy and Osmond[141] utilized CMC for improving the baking characteristics of a spray-dried baker's type of cheese. This was made by spray-drying under proper conditions an acidified milk-cream mixture containing a stabilizer blend of 3.3% CMC and 6.7% instant starch.

8. Cereals

A wholesome, nutritious breakfast is a must to start and maintain a good day at work or play.

Breakfast cereals may rely upon the thickening power and instant reconstitutability of CMC for development of desirable richness, creaminess, and full-bodied consistency. Collins and Hyldon developed "instant grits" by admixing corn grits with various emulsifiers and a rapidly hydrating polysaccharide gum (CMC) in critical amounts of aqueous solution, then drum drying the mixture into a thin sheet which is comminuted to a specific-particle-size finished product. Addition of hot or boiling water to the product yields an instant hot cereal that does not require additional cooking. The required gum usage ranged from 0.5 to 3.5% by weight of the finished corn grits. CMC type 7HOF (Hercules, Inc.) was preferred in this invention.

Addition of CMC to processed oats along with sugar and one or more flavoring ingredients makes "instant" oatmeal. Huffman and Moore[140] made an instant-type oatmeal composed of quick-cooking rolled oats and about 0.1 to 1.0% CMC. The product acquired the flavor and texture characteristics of cooked oatmeal when reconstituted with boiling water.

Matsumoto and Matsuura[208] invented a fortified artificial rice molded under pressure from a dough of flour, soybean flour, and potato flour with CMC as a binder. The mixture contained the appropriate nutrients of vitamins B_1, B_2, and calcium lactate. A subsequent process by Taninaka and Hori[209] eliminated the need for pressure treatment and produced a similar product by heating a mixture of starch and flour for 10 min at 120°C, coating with CMC, and then drying.

C. Beverages and Beverage Powders

CMC is a well-known ingredient in a wide variety of both liquid and dry mix beverage products. In these systems, CMC imparts several desirable functions including body, mouth-feel, viscosification, rapid hydration, clarity, modification of flavor perception, protein stabilization, and compatibility in mixed solvents (e.g., alcoholic beverages).

In today's marketplace, numerous food manufacturers employ CMC in instant dry mix beverages, such as fruit drinks, breakfast drinks, chocolate drinks, bar mixes, and diet drinks, for many of the functions mentioned previously.

Common[145] recognized that dry beverage powders reconstituted with water often utilize CMC as a bodying agent.

CMC is often used in conjunction with other hydrocolloids in dry mix beverages to complement and supplement their functionality. Jackman[146] discusses the use of CMC and xanthan gum together in citrus fruit juice, citrus fruit drinks or dry mix drinks containing citrus fruit solids. CMC alone in such products is an excellent bodying agent, but the gum lacks a true yield value which is necessary to suspend fruit pulp. Xanthan gum, because of its yield value, is an excellent suspending agent, but xanthan tends to cause cloud desta-bilization and pulp flocculation. In a symbiotic fashion, this invention utilizes CMC to complex with and stabilize those beverage components incompatible with xanthan, while the microbial gum suspends those components that would normally settle in the system with CMC alone.

The use of CMC to stabilize low-pH acidified milk beverages has been previously discussed (see section on milk beverages). Sirett et al.[147] proposes a dry mix composition with CMC that is suitable for rapid reconsitution with milk or water into a stable low pH acidified milk beverage. The gum is rendered readily dispersible and highly "instantized" by applying it to a readily soluble substrate. For example, the substrate, such as sugar or milk solids, is dry mixed with the gum, put in solution, then spray-dried or freeze-dried and ground. This step permits the gum to form a colloidal dispersion very quickly (more so than when not applied to the substrate) and thereby enables the gum to control the coagulation of the milk protein in the presence of the acidulants in the system.

Instant bar mixes present somewhat of a different situation as far as dry mix beverages are concerned in that a mixed solvent system (ethanol and water) is used to reconstitute the powder. The excellent tolerance of CMC to ethanol-water mixtures permits adequate body and mouthfeel to develop in these products when CMC is included in the dry mix composition. A low-viscosity, fine grind grade of CMC is a preferred choice in bar mixes for rapid hydration and good tolerance to ethanol (a nonsolvent for CMC).[86]

Hot cocoa mixes utilizing CMC offer another problem to the realm of dry mix beverages because of the thermal factor. Heat, in the form of hot water or milk, favors the solubilization of CMC in hot cocoa mixes, but it also enhances the tendency of the gum particles to stick together as clumps or "fisheyes" when dispersion (ratio of sugar to gum) is inadequate or the energy input (stirring) is weak. Medium-viscosity, uniformly substituted fine-grind cellulose gum types have been found to be particularly effective for body, creaminess, and rapid hydration without fisheye formation under the reconstitution parameters characteristic of hot cocoa mixes. High-viscosity CMC types, including fine-particle-size versions, are less effective in hot cocoa systems as the long chains dissolve more slowly and thus favor the tendency for fisheye formation.

With high-fat cocoa or ample usages of cocoa in CMC-containing hot chocolate mixes sometimes the gum and cocoa adhere together during reconstitution forming dark, undesirable floating globules. This effect may be minimized by agglomerating the hot cocoa mix which improves the dispersibility of the mix components.

Noncarbonated and carbonated beverages make use of CMC for body, clarity, and some specialized functions, like stabilization of beverage flavor emulsions. Horikoshi[148] prepared a stable lemon-flavored emulsion by treating 6.3 kg of a paste containing 3% CMC and 1% nonionic surfactant with 1.5 kg lemon oil.

In a process developed by Stewart and Falk[149] CMC is pelletized, coated with sucrose, and suspended in a saturated solution of the same coating material. The aqueous stabilizer suspensions are then sterilized at temperatures of 280 to 290°F for about 5 to 10 sec. These sterilized suspensions containing stabilizer pellets are quite useful for preparation of beverage bases. For example, addition of orange oil and a food acidulent such as malic acid to the suspension yields a highly concentrated orange-flavored beverage base.

Admixing of the base with water dissolves the sucrose coating and releases the CMC for rapid hydration. The resulting orange-flavored beverage has proper viscosity and mouthfeel. The instant stabilizer suspension also acts to maintain the beverage base oil-in-water flavor emulsion. This invention utilizing CMC facilitates the manufacture of flavored concentrated syrups for the bottling industry, generically named "throw syrups".

CMC is effective for stabilizing the foam or "head" in regular beer, and it is suitable for development of body in the newer reduced calorie beers now prevalent in the marketplace. CMC's excellent clarity and tolerance to ethanol are put to good use in beer applications. Wallerstein et al.[150] found that incorportion of CMC alkali metal salts in beer maintained highly stable foams.

CMC imparts certain gustatory qualities when added to various beverage media. Pangborn et al.[151] showed that low usages of CMC depressed the sourness of tomato juice and reduced the bitterness of coffee, making these products more desirable. The physical and oral viscosity of these products were noted to increase.

Glicksman and Farkas[152] also observed that hydrocolloids are used in special cases to mask undesirable aftertastes and to contribute toward smoother and more "blended" flavors in liquid, semisolid, and solid foods.

Low-calorie beverages (liquid or dry mix) are frequent users of CMC because of its ability to viscosify products without contributing calories to the system. When artificial sweeteners are used in sugarless beverages, CMC imparts a syrupy mouthfeel that is normally supplied by considerable concentrations of sugar. Brenner[153] originally suggested the use of CMC as a bodying agent in sugarless beverages containing sodium cyclohexylsulfamate instead of sucrose.

D. Pet Foods

Mankind's fond devotion and preoccupation with pets for companionship has been known since ancient times. The old adage "man's best friend" aptly describes this special relationship. Man's concern for animal friends is vividly reflected in the quality and nutritional soundness of today's pet foods. Just as human food has grown and diversified throughout the years, pet foods have become an industry within the food industry. Several companies specialize only in this single category of food products.

Current pet foods are to a large extent modeled after human food products in that they must possess the correct color, flavor, odor, eye appeal, nutritional balance, packaging, and price. After all, it is the consumer, not the pet, who purchases the product. The physical stability and texture of pet foods also influence the purchase appeal of these products. A separated product is of no concern to the pet, but may cause the consumer to switch brands. It is hydrocolloids, like CMC, that play a vital role in the physical stability, texture, and "consumer appeal" of pet foods.

Within the class of pet foods, there are three basic subdivisions — dry, semimoist, and moist pet foods. Moist pet food is in total a canned product — the most common and oldest known form of packaged animal meals. Meatball-in-gravy or chopped meat types are typical varieties found in the U.S. marketplace. Gelled, canned pet foods are popular in Europe, but are relatively uncommon in the U.S.

CMC is utilized in canned pet foods for several functions. It serves as a binder to prevent water exudation and helps stabilize meat emulsions which deters fat capping or collagen separation during processing. It helps regulate product consistency which facilitates a more uniform filling operation prior to retorting.

In meatball-in-gravy type products, CMC is used as a physical binder in the meatballs to maintain structural integrity. CMC is also used to thicken the gravy portion of this product which helps reduce soakback of the gravy into the meatball.

In chopped-meat canned pet foods with gravy, CMC thickens the gravy and provides some "cling" to the meat chunks.

Generally, high-viscosity CMC types are preferred for canned pet food items since pet food manufacturers insist on maximum viscosity per pound of gum. Typically, canned pet foods are retorted under severe processing temperatures, such as 260°F for 60 min or more. Since no gum is perfectly retort stable, some degradation of CMC would be anticipated under such extreme processing conditions.

Moist pet foods may rely on citrated blood for color, flavor, and high nutritional input. Bernotavicz[154] describes a rather novel invention whereby blood chunks are prepared by combining blood and at least one gum with an optional protein source (soy), heating the mixture and comminuting the solidified mass to a desired particle size. CMC was found to produce a firm blood clot when added at a preferred level ranging from 0.75 to 1.25%. Blood chunks made by this process are suitable for use in canned pet food because they do not disintegrate under canning conditions and retain their structure. Thus, realistic color, flavor, and texture may be imparted to canned moist pet food.

Semimoist pet foods are a more recent development in animal foods. These products contain enough water to achieve a meaty consistency and sufficient amounts of water-soluble solute to raise the osmotic pressure to a level where microbiological decomposition cannot occur. Hence, these products are shelf stable at ambient temperatures.

CMC functions in semimoist pet food to bind moisture, to maintain structural integrity, to impart pliability, to aid the extrusion process, and to improve the cosmetic appearance of the product.

Bone and Shannon[155] describe a process for making a pet food with an abrasive dry component for cleaning the animal's teeth and a soft semimoist component for palatability. The soft component is prepared using a premix of propylene glycol, sucrose, CMC, and water to achieve the necessary plasticizing/humectant effect and promote a meat-like consistency.

Burkwall[156] devised a combination simulated egg and meat semimoist pet food. The invention specifies the use of various ''edible water-absorbing hydrocolloids'' like CMC in the simulated egg portion of the product. Both the simulated egg and meat portions are extruded into hamburger-type strands which may be packaged loose or molded.

Simulated cheese and meat or multimeat semimoist pet food products are marketed by manufacturers which contain small amounts of CMC for various textural and cosmetic function.

Dry pet foods which are available in many shapes and sizes occupy a major market share of comestible pet products. These items are simple and easy to store, require the least packaging, and are very shelf stable because of the low moisture content. The product is commonly in the form of a pellet or kibble which is produced by a pressurized extrusion of a farinaceous material such as corn. With all of the inherent advantages of dry pet foods, it still remains difficult to feed pet food in the dry form due to palatability problems.

Consumers may overcome palability drawbacks in dry pet food by simply adding water. Pet food manufacturers have taken a further step by coating or ''dusting'' various dry ingredients, such as flavor, color, or gums, onto tallow coated dry pet food so that addition of water facilitates formation of a gravy with desirable appearance, taste, aroma, and consistency. CMC, because of its rapid hydration and viscosifying power, is a popular component of dry, gravy-forming pet foods. A well-known dry pet product in the marketplace with a gravy-forming capability is Gravy Train® (General Foods Corp.).

A relatively recent application for CMC is in pelleted animal feed.[157] The primary function of the gum is to improve pellet durability. CMC serves as a physical binder in the pellets to maintain structural integrity and prevent the accumulation of fines in feed packages during handling and transit. Pellet durability is an important quality standard to feed manufacturers.

In pelleting operations, CMC functions as a lubricant. It allows pellet mills to be operated at higher temperatures and higher moisture levels without plugging. Increased production rates and lower energy consumption are realized with addition of the gum to pelleted products. CMC-7LF at 0.5 to 1.5 lb/ton is recommended for pelleted feed products.[157] At increased usage, 10 to 20 lb/ton, CMC is recommended for use in aquatic feeds to prevent too rapid a breakup of the feed particle in an aqueous environment and thereby facilitates consumption.

In a related application, Dannelly[158] developed rumen stable animal pellets which utilize CMC. These pellets consist of a matrix containing a nutrient, e.g., 500 g of methionine, 20 g sucrose, and 10 g of CMC, to which water (135 g) is added to make an extrudable wet powder. Dry pellets are then obtained by adding a 50/50 mixture of sucrose/CMC to the wet pellets and tumbling with a hot air input. A hydrophobic coating comprised of fats, aluminum flake, and talc is then applied. The pH of the pellet core is a desired 5.68 or greater.

These pellets are designed to be orally administered to ruminants. The materials used in these specialized pellets create a protective effect against the digestive conditions of the rumen (pH 5.4 to 5.5), and a subsequent rapid breakdown of the mass occurs in the abomasum/intestines at pH 2.9. Thus a far more effective means of administering essential amino acids to ruminants is obtained. Normally the rumen flora would metabolize essential amino acids to a certain extent which limits growth efficiency of the ruminant.

E. Salad Dressings, Condiments, Gravies, and Syrups
1. Salad Dressings
CMC is permitted as an optional ingredient (stabilizers and thickeners) for salad dressings, including French dressings, under the standards of identity for food products of this type (CFR Title 21, Part 169). Even so, most modern cold process salad dressings have turned to other hydrocolloid types which more closely fit the stability needs of such pourable dressings with specialized properties, e.g., a yield point.

An exception for the use of CMC in salad dressings is instant dry mix dressing contained in a packet. These products are designed as convenience items to which water, milk, and vinegar are added in a cruet and subsequently the mixture is shaken to the desired consistency. CMC, because of its rapid hydration, thickening ability, and cost-effectiveness, finds use in this particular variety of dressing.

2. Condiments
Anderson et al.[159] found that combinations of CMC (0.25 to 0.75%) and carrageenan (0.15 to 0.45%) were effective in reducing syrup separation from sweet-pickle relish during storage. In addition to improved drain-weight values, relishes prepared with these gums appeared less dry and brighter in color compared to controls. Improved relish appearance was attributed to the formation of a thicker film of syrup surrounding each relish particle.

3. Gravies
Meat and poultry pot pies and baked pies can be improved by the addition of CMC as a thickening agent for the gravy. The gum provides improved gravy appearance (sheen) and firm consistency so that development of a watery exudate after baking is minimized.

4. Syrups
Table syrups, particularly pancake syrups, have gained widespread popularity among consumers in recent years. No stack of hotcakes would be complete or palatable without ample addition of a rich, thick, flavorful syrup. Syrups are available in a variety of colors, flavors, (maple, butterscotch, chocolate, blueberry, or strawberry), sugar contents, and consistencies. To maintain this latter characteristic, CMC is commonly employed in syrups as a flow control agent. The clarity, noncaloric value, thickening power, and compatibility with sugar are attributes of CMC which make it particularly well suited for syrup applications. Conventional high brix syrups (65% or more soluble solids) generally utilize small amounts of lower molecular weight types of CMC to adjust viscosity to a desired value.

Tremendous interest has arisen during this decade in the marketing of reduced calorie table syrups. These products have reduced sugar content (less than 65% soluble solids) and build back viscosity (without calories) by the addition of CMC. Generally, medium viscosity CMC types with a high or uniform DS are preferred in these products to achieve smooth flow characteristics.

High-viscosity, randomly substituted CMC types are not advocated for table syrups as they readily impart a grainy, slimy texture because of their thixotropic nature and high DP (degree of polymerization).

Dietetic syrups employ CMC and artificial sweeteners to yield viscous syrups with no caloric value. Since higher amounts of CMC are required by these systems compared to syrups containing sugar, smooth flow and clarity are mandatory requirements which CMC readily fulfills. High-DS, uniformly substituted, medium-viscosity CMC grades are also preferred in these systems.[160]

CMC is effective in preventing sugar crystallization in syrups and thus controls the formation of undesirable white sugar deposits on the neck or screw cap of polyolefin containers now used to package syrup products.

Another function of CMC in syrups is the prevention of soakback which causes sogginess in products such as pancakes. Thin watery syrups readily absorb into pancakes to give a mushy consistency. The consumer also has the tendency to keep pouring on more syrup as soakback progresses. This becomes expensive for the consumer (especially with institutional syrups in restaurants). By thickening syrups with CMC, the tendency for soakback into pancakes is retarded and a cost savings may be realized.

F. Confectionery

In confectionery products, the inhibition of sugar crystallization is mandatory in order to achieve the desired properties of gloss, body, prolonged shelf life and smooth creamy texture. Mechanically, it is known that the process of sugar crystallization involves the dissolution of small sugar crystals simultaneous with the growth of larger crystals. Interference in this process is made by ingredients known as ''doctors'' which include materials such as corn syrup, invert sugar, milk solids, fats, emulsifiers, and various hydrocolloids.

Desmarais and Ganz[44] investigated the efficiency of CMC as a doctor alone or in combination with other doctors in supersaturated sugar solutions and fondants. This work revealed that all of the doctors evaluated (corn syrup, invert sugar, and CMC) were effective in inhibiting sugar crystal growth. Low-viscosity CMC was determined to be more effective than other doctors, including high-viscosity CMC, in controlling sugar crystal growth as shown by a high percentage of small crystals over a 4-month storage period.

In an all sucrose fondant, which normally has a short shelf life, Desmarais and Ganz[44] lengthened shelf life by incorporating low-viscosity CMC. Photomicrographs of fondants creamed at 110°F and aged for 2 months show significant reduction in sugar crystal size (4 to 18 μm) with 0.25% low viscosity CMC compared to a control (4 to 80 μm) without CMC. It was observed that the CMC was more effective when added prior to or during boiling rather than just before creaming. When a small amount of corn syrup (3%) was added to these fondants, CMC was found to broaden the temperature range (from about 110°F to a maximum of 150°F) at which the fondant is creamed on a slab-type beater.

Andersen[161] developed a sugar-free confectionery material based on xylitol and sorbitol. The invention specifies the addition of a sugar-free swelling agent, such as CMC, at 20 to 50 wt% to impart required plasticity and cohesion so that an attractive chewy sweet may be readily formed. Furthermore, the tendency of xylitol to recrystallize is suppressed by the hydrocolloid.

In licorice-type confections with soft white or colored centers, it is reported that small amounts of CMC in the centers improve mouthfeel and extend shelf life by preventing the centers from drying out.

Sugar confections are often coated with lake colors and lacquers by a panning technique. The process time is lengthy and numerous coats may have to be applied to achieve the desired appearance. The utilization of CMC-7LF as a binding agent in syrup containing food lake colors decreases processing time and reduces the number of coats required.

G. Dietetic Foods

In various feeding studies, CMC has been determined to be physiologically inert since it is passed through the intestinal tract unchanged and as such is noncaloric. This property, in addition to the gum's pronounced bulking ability, makes CMC useful in the field of dietetic or low-calorie foods.

Ferguson[162] described an appetite-reducing food consisting of CMC and tartaric acid (1:4). This composition dissolved slowly in the mouth and coated the taste buds such that keenness of taste was masked. Presumably, one's desire to consume high-calorie foods was appeased.

Glicksman cites the contribution of Pfizer & Co. in 1962 who entered the dietary food field with a low-calorie cream-filled biscuit containing the "bulking agent sodium carboxymethylcellulose to allay hunger and at the same time satisfy the desire for solid foods."[10] These biscuits, which sold under the name of Limmits®, each contained 175 calories. The standard diet consisted of four biscuits per day plus two glasses of whole or skimmed milk together with one normal low-calorie meal. The function of the 2.0% sodium carboxymethylcellulose, according to the promotional literature (Charles Pfizer & Co., 1962), was to supply the bulk, "thereby helping to appease the appetite, counteracting a frequent cause of failure in weight control programs."[186] On contact with the stomach digestive juices, the CMC expands slowly to produce a bulk effect. The biscuit is supplemented by sufficient fat, which is digested slowly and gives the diet "staying power". This product was apparently not commercially successful, since it was withdrawn after a short time.

Several low-calorie or dietetic food products are based on hydrocolloid gels to supply the necessary bulk and body usually imparted by high sugar concentrations. Dietetic jams, jellies, and dessert gels utilize pectin, carrageenan, or alginates to obtain a gelled consistency, but the complete lack of sugar is difficult to overcome and these products tend to suffer inherent problems, such as weeping. Addition of a small amount of CMC (e.g., 0.25% 7HOF) to such gels provides water binding and decreases syneresis. Concurrently, CMC maintains the noncaloric requirements for these products.

The functional contributions of CMC to low-calorie or dietetic syrups, beverages, and confections have been previously mentioned in the respective sections for each of these products.

Other related low-calorie applications for CMC include low-calorie bread, pharmaceutical wafers, low-calorie cakes, low-calorie canned fruits, and low-calorie spreads, creams, or sauces. Nijhoff[30] describes the preparation of low-calorie dressings, spreads, creams, and a strawberry dessert made by application of shear to CMC in a DS range of 0.1 to 0.6. A spread similar in unctuous character to margarine or butter with only half of the calorie content is cited in the disclosure. Pastry creams, either chocolate or vanilla, with a marked reduction in caloric content compared to conventional high-fat creams were prepared by similar procedures. These creams were soft, yet shape retaining, pleasant in taste and of a consistency suitable for use in a variety of soft bakery goods.

Rispoli et al.[163] describes an oil-replacement composition which is oil-free, but posseses the mouthfeel, texture, and lubricity of oil without contributing the full caloric impact of oil. The composition is suggested for use as a partial or full replacement for oil in salad dressings (e.g., mayonnaise), desserts (e.g., ice cream), margarine, etc. To prepare the oil replacement, a protein and CMC are blended, hydrated, and whipped. Secondly, a modified starch and an acid are mixed, heated (which swells the starch), and cooled. Part A (protein-CMC) and Part B (starch-acid) are then mixed together with continuous agitation to obtain the oil-replacement composition. The interaction of CMC with protein at acidic pH (3 to 6) is cited as an important factor in the invention.

H. Processed Foods

Napoleon first recognized that an "army marches on its stomach" and offered a prize of 12,000 francs to anyone who could devise a better means of keeping food fresh longer. Thus he hoped to improve battlefield logistics. Although Napoleon met his defeat at Waterloo, his ultimate triumph came about in the food industry. In 1811, it was the challenge and incentive of Napoleon's decree to make food less perishable that inspired Nicholas Appert to seal foods in containers and subsequently heat the enclosed system to make the food within edible and safe for a prolonged period. Appert's discovery, which we recognize today as canning, earmarked the beginning of the food processing industry which has impacted immensely on the progression of modern man.

Most food items undergo some type of processing before they reach the consumer in today's fast paced society. Foods may be frozen, dehydrated, canned, precooked, deep-fat fried, freeze-dried, blanched, extruded, and the list goes on. All of these procedures serve to improve or extend the keeping qualities of food, provide the highly desired aspect of convenience, and satisfy the decree set forth by Napoleon.

CMC plays a prominent role in the processing of foods in that it thickens, stabilizes, and helps alleviate some of the wear and tear imparted to foods during processing. Previous sections of this chapter have frequently touched upon the role of CMC in processed foods and some additional related applications will be elaborated upon here.

1. Frozen Foods

The incorporation of CMC into ice cream and related frozen desserts to control ice crystal growth and prevent other quality defects was discussed earlier. This functionality has been adapted to other types of frozen foods.

Fruits and vegetables are commonly frozen to insure stability and keeping qualities during shipment. However, plant tissues are subject to structural damage during freezing and upon thawing significant weight loss from water exudation may occur. CMC has been used in frozen fruits and vegetables to control ice crystal growth damage to tissues and maintain adequate drain weights. Akuta and Koda[164] used CMC to preserve the organoleptic qualities and ascorbic acid content of frozen strawberries. Burt[165] developed a fruit preservation method which consisted of drawing a vacuum on apple slices, blanching the slices in a thickened sugar solution containing 0.2 to 1.0% CMC to maintain the firmness and natural color of the apple slices, and then preserving the blanched slices in the sugar solution by freezing.

CMC is used to improve texture and prevent syneresis in frozen pie fillings, fruit tarts, fruit laden pastries, and cheesecakes.

Mason[177] describes a fruit preservation process whereby a variety of fruit products (diced, sliced, or in sections) are treated with dry CMC (0.25 to 0.5%, based on fruit weight) by tumbling or other suitable mixing technique so that gum particles are admitted into the open pore structure of the fruit piece. Once this is accomplished, the gum swells, creates a pore blockage and moisture loss or "bleeding" from the fruit piece is diminished. This clogging effect helps the fruit piece retain its original shape and texture with no undesirable loss in firmness during heat processing or freezing. This invention "locks in" flavor which would normally be lost as the juices exude and degradative reactions such as "browning" are reduced since exposure of the fruit piece to atmospheric oxygen is minimized.

Latham and Seeley[166] describe the use of CMC as a thickening agent and suspending aid in a batter composition used to make cooked frozen omelets and other frozen egg products. In addition to viscosity control, the gum prevents water exudation from the batter when it is thawed for use and assists in volume retention when the omelet is cooked. Seeley and Seeley[167] developed a cholesterol-free egg product which may be conveniently made into frozen egg patties with improved cooking tolerance. The improved cooking tolerance of the formulation results from the addition of carrageenan which provides a gel-strengthening effect.

The invention further specifies the addition of a combination of CMC types (7LF and 7H3SF) to improve the gel strength of the carrageenan in the formulation as the product cools just after cooking.

Shemwell and Stadelman[168] showed the effectiveness of 1% or less CMC in preventing undesirable weepage from frozen spinach soufflés upon baking.

Battered and breaded frozen chicken patties and chicken sticks have become increasingly popular for consumption at home or in fast-food restaurants as chicken burgers or sandwiches.

To increase the moisture retention in these products after they are reheated by microwave or conventional oven procedures, manufacturers incorporate, by injection or admixture, a solution of CMC into the meat matrix prior to freezing. This gives enhanced moisture binding and a juicier texture. This technology is being expanded to other related meat items in the current marketplace.

2. Dehydrated Foods

Certain dehydrated food such as fruit and vegetable powders or dry soup powders rely upon the addition of CMC prior to dehydration to provide ease of reconstitution and restoration of palatable texture upon reconstitution. Perech[169] used CMC in the conversion of fruit and vegetable juices to dry powders by dehydration so that a thickened product evolved upon reconstitution. Bogin and Feick[170] described the utilization of CMC, water, and fatty materials, such as chicken fat, to form dry, stable, nongreasy solid food materials by dehydration. This procedure allowed a fatty substance to be "reconstituted" when used in dry packaged items such as soup mixes and the fat was protected from degradation by oxidative rancidity.

Galle et al.[171] prepared simulated nutmeat products which utilized CMC to stabilize an oil-in-water emulsion that is subsequently spray-dried. These particles are then molded using a tableting machine into pieces having the surface configuration of a nutmeat. Addition of various dry flavors to the spray-dried emulsion prior to molding produced different simulated nutmeats — walnut, pecan, etc.

3. Canned Foods

Several of the functional attributes of CMC in canned food products were previously described in the discussion on canned pet foods.

CMC was shown to be particularly effective for prevention of liquid separation in fresh fish sterilized at 110 to 115°C in closed containers.[172] Kojima et al.[173] cited the ability of CMC to reinforce gel strength in canned agar fruit gels. Agar gels were strengthened with 0.1% CMC and were packed with fruits and heavy sugar syrup in canned mitsumane. These products were stable after 80 days at 30°C and possessed firmer texture than agar controls without CMC.

CMC effectively controls liquid separation in canned pie fillings and canned pudding products.

4. Freeze-Dried Products

Gejl-Hansen and Flink studied the incorporation of CMC into the matrix of a freeze-dried oil-in-water emulsion by various microscopic methods including SEM (scanning electron microscopy).[174] CMC at 0.5% w/w was observed to effectively "encapsulate" the oil component of the emulsion and thus would influence the oil stability and rehydratable properties of the emulsion. This functionality would aid in the design of improved dehydrated emulsion-based engineered foods.

5. Processed Meats

The functional properties of CMC are finding new and novel applications in processed meat items. Recently developed chicken franks and turkey rolls utilize small amounts of CMC to help immobilize the aqueous phase of the meat emulsion, prevent weepage (especially in vacuum packaging), and enhance the shelf life of these products. Sausage and other meat analogs have appeared in the marketplace. These products rely upon nonmeat components such as soy protein to recreate a meaty texture without the calories, cholesterol, and saturated fats which are often scrutinized in conventional sausage. In analogs, a solution of CMC may be used in conjunction with soy protein chunks or granules (e. g., TVP). The gum solution swells the granules to a meat-like consistency and helps retain the moisture in the granule during processing. Without CMC, soy chunks develop a mushy, nonmeaty consistency upon reconstitution which is undesirable.

Bomstein[175] developed a texturizing aid (fibrous material) for extension of meatballs, luncheon meats and other related meat products. The fibrous product was prepared by acidifying a chitosan solution, then combining it with a CMC solution which yielded a fibrous precipitate that was collected on a sieve. The fibrous product so obtained was white, bland, and possessed a texture closely resembling that of meat.

Pader et al.[176] describes the use of CMC (0.19%) and guar gum (0.09%) in the stabilization of an emulsion used in the preparation of a soy-protein-based spread, resembling a meat spread, that is used on sandwiches or crackers. The emulsion is used at an "edible outer additive" which coats the inner matrix of protein flakes preventing the binding of such flakes during heat processing and imparts spreadability to the entire composition.

Mechanical deboning of chicken and other types of poultry has been investigated by many workers and various techniques are available in the patent literature. However, deboned chicken meat tends to become mushy and pulpy or undergo rapid moisture loss once it is removed from the original framework. CMC is added to mechanically deboned chicken meat to help maintain a meaty texture and act as a moisture binder to prevent "pooling." This deleterious effect reduces uniformity and makes the deboned meat less desirable for use in subsequent food products.

It has been reported that application of CMC in solution between layers or slabs of bacon or bacon analogs reduces the tendency of the layers to fuse together during storage and facilitates ease of separation at the time of use.

I. Food Preservation Applications
1. Preservative Food Coatings

CMC, like many other cellulose based polymers, has the ability to form films, and this property has been used to good advantage in the protection of perishable foodstuffs from bacterial and chemical degradation.

Glicksman[10] has provided an excellent review of pertinent food preservative coating applications using CMC which were known prior to his publication.

Cornwell[178] used CMC in conjunction with antimycotic agents to coat meats, fruits, and vegetables and thus protect them from bacteria and molds. In a similar way, eggs were also protected by a film of CMC containing a preservative (Algemene Kunstzijde Unie N. V., 1952).[179] Improvements in this technique were made by Vale,[180] who formulated a solution consisting of about 1.0% of an alkyl quaternary ammonium halide, 0.1% CMC, and glycerol as an optional plasticizer. This composition was effective as a temporary preservative coating for fruits and vegetables for a period of about 6 weeks. (i.e., between picking and canning) and could be removed by washing with water at final processing.

Shinn and Childs[181] used CMC in a meat-coating preparation comprising an emulsion of 30 to 60% fat, 30 to 60% water, 2 to 12% edible gelatin, and 0.05 to 3% CMC. Brown et al.[182] dissolved a dry mix of 2 parts sorbic acid and 3 parts medium-viscosity CMC in 95 parts water to form a dip or spray that formed an antifungal coating on food products. In a similar manner, Nikkila et al.[183] used CMC to thicken a preservative solution containing soluble phosphates that was used to treat fresh food prior to or during freezing.

More recent applications for CMC as a functional coating are presented as follows. Ghosh et al.[184] developed an effective fungistatic wrapper made by coating a grease-proof paper with an aqueous dispersion of sorbic acid in 2% CMC solution. The shelf life of foodstuffs such as bread can be prolonged by wrapping them in the treated paper and then enclosing in a polyethylene bag.

Hutchinson and Swanson[185] describe the coating of edible fibrous cellulose with a CMC solution containing polyhydric alcohol and then drying the mixture to provide a free-flowing product. This coating procedure imparts smooth texture and ease of palatability to the normally dry, gritty taste and texture of the cellulose fibers. Aside from the noncaloric effect, this invention facilitates the addition of dietary fiber in Western Man's diet which is believed to have an important role in the reduction of various diseases and ailments.

2. Encapsulation

The film-forming ability of CMC permits its use as an encapsulating agent for fats, oils, and other materials that degrade readily in the presence of molecular oxygen. The low oxygen transmission property of CMC films is highly functional in this application.[187]

The use of CMC in flavor encapsulation is quite limited. Katayama et al.[188] describes an encapsulation procedure whereby two oppositely charged polymers separate from aqueous solution as a coacervated complex which coats oil droplets and is hardened via crosslinking to form capsules. Gelatin and gum arabic are the preferred colloids in this procedure. During the hardening pretreatment of the coacervate containing an oily liquid, a shock phenomenon occurs when the pH of the system is adjusted near the isoelectric point of gelatin, causing a rapid increase in viscosity which interferes with formation of singular microcapsules. CMC is specified in this invention as a shock-preventing agent. CMC prevents the rise in viscosity characteristic of the shock phenomenon; however, the exact mechanism of the prevention is not well understood. The known interaction of CMC and gelatin is undoubtedly involved in the mechanism.

Other workers cite the use of CMC in encapsulation for thickening aqueous phases, as a filler, as a carrier, as an emulsifying aid, or for enveloping materials such as vitamins; however, other gums may function similarly in these specific roles.[189-191]

3. Fish Preservation

Seafoods are perhaps the most difficult of edible materials to keep fresh for any significant time period. It has been shown that antibiotic treatment of fish, such as with chlortetracycline (CTC), permits fishing vessels to stay at sea longer with less occurrence of the catch spoiling.

The CTC is applied to fish at about 1 ppm by incorporation into ice blocks used to ice down the fish. However, uneven distribution of CTC in ice blocks tends to be a problem. Boyd et al.[192] found that addition of CMC or carrageenan produced a more uniform distribution of antibiotics in ice blocks prepared by conventional slow freezing procedures.

A less scientific method of fish preservation involves the freezing of fish in saline solutions thickened with CMC.[193]

4. Tablet Coating

In a food-related use, CMC is utilized in various tablet coating applications because of its noncaloric character, its film-forming ability and its functionality as a physical binder. Spradling[194,195] used CMC in a sugar syrup to give a more elastic and chip-resistant subcoating for medicinal tablets. Claims were made that fewer coatings were required and ease of swallowing was improved.

Doerr et al.[196] explored the use of CMC and HEC (hydroxyethylcellulose) as alternatives for sugar coatings on medicinal tablets. These newer coating materials significantly reduced processing time in comparison with much longer times required by sugar coatings, thus reducing labor costs. The cellulosic coatings withstood infrared heat and a durability test better than various commercial coatings compared to them.

Long[197] described the utilization of a combination of CMC and shellac to prepare coatings for tablets that can rapidly disintegrate in the stomach, even after aging, compared to shellac alone or other coating materials. This coating composition has improved smoothness, luster, and chip resistance.

Christenson and Dale[198] developed a sustained-release tablet which employed CMC. By combining a preponderance of CMC and a drug into a pressed tablet, the composition becomes a swollen mass (a large fisheye) in the stomach juices which dissolves slowly to give a sustained release of the drug. This invention makes it possible to maintain the action of the drug during the night while the patient is asleep as well as during the day when it may be inconvenient to take frequent dosages of medication.

J. Instant and Gelled Dessert Products

Fanciful and flavorful desserts are the perfect finishing touch to any elegant dining engagement. Just as these items conclude the dining experience, it is perhaps fitting and proper to consummate this section of the chapter with a look at the use of CMC in dessert products. Since the application of CMC to baked and frozen desserts has previously been covered, this discussion will focus on CMC functionality in instant, gelled, and related dessert items.

Instant dessert mixes, including puddings, pie fillings, and mousses, traditionally contain a preponderance of starch along with fats, emulsifiers, and phosphates to develop body and proper set. Certain manufacturers add a small amount of CMC to these products to combat syneresis and shorten mix time due to the rapid hydrating ability of CMC in cold milk or water.

Instant and processed whipped toppings are popular adornments for many desserts, including puddings, pies, mousses, ice cream, and strawberry shortcake. Irish coffee, although not a true dessert, is commonly served with an appropriate whipped topping.

CMC is an effective stabilizer in whipped toppings where it inhibits weepage, thickens, aids overrun, enhances the foam structure, provides ease of reconstitution, and gives acceptable whipping time.

A popular stabilizer combination for instant whipped toppings prepared with milk or water includes CMC and another cellulosic polymer, Klucel® (hydroxypropylcellulose). The CMC provides the all-important moisture binding property while the Klucel®, which is surface active, helps aerate the system during mixing. Both gums thicken, hydrate rapidly, and reduce whipping time in this application. High-DS CMC types are preferred in instant whipped toppings for maximum moisture binding. Another stabilizer system encountered in instant whipped toppings is a combination of CMC and a lambda-type carrageenan.

Processed, packaged whipped toppings demand the same stabilizer functionality as instant whipped toppings with the added requirements of freeze-thaw tolerance and longer shelf life. CMC is effectively used to meet such requirements.

Jonas[199] developed a multipurpose whipped dessert which may be consumed in the frozen state as an ice-cream-type dessert or in the thawed state as a whipped topping. The edible whipped dessert is comprised of an aerated admixture of a topping emulsion and a protein emulsion. A combination of CMC and sodium alginate is a preferred stabilizer system in this product which provides desired stiffness, foam structure, freeze-thaw stability, resistance to syneresis, and facilitates the multipurpose functionality.

Strums and Jonas[200] describe a low-fat whipped dessert composition which is aerated by passing through an Oakes mixer to develop the foam structure and make the product suitable for freezing. A stabilizing agent is required in the product which consists of a combination of CMC, sodium alginate, and tricalcium phosphate at 0.10 to 0.15% by weight in the finished product.

Gilmore and Miller[201] disclose a low-calorie whipped topping mix that is in spray-dried form. The composition contains a whipping agent so that an aerated state may easily be achieved upon addition of water or milk with subsequent agitation. CMC is a preferred stabilizing gum in the mix which offers body, moisture binding properties, and freeze-thaw tolerance to the finished product.

Pader and Gershon[202] describe an aerosol topping dispensable from a pressurized container in the form of an aqueous emulsion. A suggested composition in this patent contains 30% fat, 0.6% emulsifier, 10.5% sucrose, 0.3% sodium caseinate, 0.2% CMC and the remainder being minor ingredients and water.

Gelled dessert products have long intrigued consumers because of the wide variety of textures available and the opportunity to "develop the gelled product, magically, in one's own kitchen." Various gelling hydrocolloids, including gelatin, pectin, alginate, and carrageenan are commonly used to prepare gelled dessert products. Although not a true gelling hydrocolloid, CMC is a functional component in various gelled desserts where it may firm, modify, or supplement the gelled texture and correct deficiencies arising from the gelling agent, such as syneresis.

Igoe[203] prepared a stable low pH milk gel using a blend of CMC and gelatin. The CMC functions to protect the milk protein in the acidic environment and promotes thickening. The gelatin provides the gelled consistency. Pedersen[42] describes a process for preparing a gelled, fruit-flavored, sour milk product whereby CMC or high methoxyl pectin is used to stabilize an acidified milk preparation (pH 3.8 to 4.2) during pasteurization to avoid coagulation of milk protein, and the product gels during postpasteurization cooling from inclusion of a calcium-reactive low methoxyl pectin. CMC type 7MF was preferred in this invention. The dessert possessed a smooth, soft gelled texture at 5°C whereas other gelling agents would become too firm and unpalatable at refrigerator temperatures.

Igoe[204] describes another means of preparing an acidified milk gel using a trigum system of CMC-xanthan-locust bean gum. Again, the CMC functions to protect the milk protein during heat processing in the presence of fruit-derived acidulents. The synergistic interaction between xanthan and locust bean gum forms the gel. These gels may be prepared by a one-step process where all the gums are added simultaneously or by a two-step process where CMC is first interacted with acidified milk, then the gelling gum combination is included.

Water gel desserts prepared with gelling agents such as carrageenan, gelatin, agar, and alginates must be clear, stable, possess a pleasing texture, have good mouth meltdown, and in some cases, be devoid of calories. Syneresis is a common problem in water gels. It is a favorite trick of product developers to supplement water gels with CMC where its excellent moisture binding property counteracts the development of undesirable syneresis. With brittle carrageenan water gels, CMC modifies the gel by imparting more elasticity without altering desired clarity. This modification is a more cost-effective means of carrageenan gel alteration compared to the use of clarified locust bean gum which is becoming increasingly expensive.

In dietetic dessert gels, CMC may be used to impart more firmness to the gel, prevent syneresis, and maintain the desired noncaloric requirements.

Freeze-thaw stable products are ongoing obsessions in the R & D departments of many food companies. The addition of a small percentage of CMC to water gel desserts imparts improved freeze-thaw stability to these products. For low pH water gel desserts, acid tolerant CMC type 7HOF at 0.25% is suggested for improved freeze-thaw performance.[205]

Glicksman[206] used CMC in an alginate dessert gel designed to simulate a gelatin-type dessert or aspic that could be frozen and thawed without textural degradation or other quality deterioration.

UHT processed, aseptically packaged, ready-to-serve desserts are a relatively new technological development of European origin. These products employ some rather unique thixotropic carrageenans that allow a gel to be pumped as a fluid below the gelation temperature, and then reformed upon shear removal. Thus a milk gel may be processed through UHT equipment, filled cold, and yield an intact reformed gelled structure.

Starch is commonly used as a bodying agent in these UHT products. It has been found that CMC can effectively replace, partially or totally, the starch used in conjunction with carrageenan in gelled UHT milk desserts processed by indirect heating (280°F with 4-sec hold) and subsequently cold-filled at 50 to 60°F.[207] CMC type 7HF at very low usage (0.07 to 0.1%) gives improved body without heavy starch consistency or mouthfeel in UHT milk desserts. Also, very effective prevention of syneresis is achieved, which prolongs shelf appearance of these milk dessert products.

K. Miscellaneous

Numerous miscellaneous food and food related uses for CMC are known. Some of the more interesting of these applications are presented below.

Microcrystalline cellulose sold under the trade name Avicel® is prepared by acid hydrolysis of α-cellulose under special reaction conditions.[210,211] The acid insoluble crystalline residue is washed, separated, and dried to yield the microcrystalline cellulose. Avicel® has excellent bulking properties, is noncaloric, and is quite stable to heat and/or acid, which makes possible some rather novel food applications. The material is commercially available as a fine white powder that is inert to organic solvents, fats, or oils, and is nonfibrous. Although Avicel® has water absorptive properties, it is essentially water insoluble. Some types of Avicel® require more energy than others to disperse properly. Therefore, these types of Avicel® incorporate about 10 to 15% (by weight) CMC which serves a dual function:

1.   Barrier — During the drying process (spray- or fluidized bed drying) in the manufacture of Avicel® CMC acts as a barrier which prevents bonding (hydrogen) of particles to each other so that a powder results.
2.   Aqueous dispersant — When Avicel® is added to aqueous systems or foods, integral CMC aids dispersion. As the CMC solublizes, the polymer chains unwind and help push the colloidal Avicel® particles apart.

The ability of CMC to assist in the dispersion of other materials is not limited to Avicel®. McGinley and Zuban[212] developed a method of dispersing "difficult" oleaginous materials, such as emulsifiers or fats, in cold water quickly and conveniently. The oleaginous material is absorbed into a matrix of 85 to 95 parts beta-1,4-glucan (preferably Avicel®) and 5 to 15 parts CMC. The matrix is combined with a critical amount of water, mixed into a paste and subsequently dried to produce free-flowing powder that expands rapidly when reconstituted to effectively disperse the oleaginous substance.

Bishop[213] developed a solid tea cube bound with CMC which could be shelf-stored in a package and readily reconstituted with hot water to conveniently prepare brewed tea. The addition of 2% CMC solution to loose tea leaf was made and the mixture was compressed into a solid matrix (cube).

More so than other food-related CMC applications, toothpaste merits a short discussion herein because of its postconsumption role. Dicalcium-phosphate-based toothpaste contains about 25% moisture. Here, CMC functions to thicken the aqueous phase, control syneresis, and impart the important property of "standup" to the toothpaste. The choice of what CMC type to use in dicalcium-phosphate-based toothpaste depends on the properties one desires.[214] Lower-DS CMC types (e.g., 7MF) are more thixotropic and promote better "standup". Higher-DS CMC grades (e.g., 12M31P) give improved gloss and smoothness, but provide less structure. Silica-gel-based toothpaste is characterized by a lower moisture content (<20%). In these products, the silica imparts nearly all of the thickening, so CMC is utilized to control syneresis.

VI. REGULATORY STATUS

A. FDA and FCC Status

In the U.S. purified sodium carboxymethylcellulose (cellulose gum) suitable for food use is categorized by the FDA (Food and Drug Administration) as GRAS — generally recognized as safe. This standard is set forth in the U.S. Code of Federal Regulations, Title 21, Section 182.1745.

The FDA defines the direct food additive as the sodium salt of carboxymethylcellulose, not less than 99.5% on a dry-weight basis, with a maximum substitution of 0.95 carboxymethyl groups per anhydroglucose unit, and with a minimum viscosity of 25 cps in a 2% (by weight) aqueous solution at 25°C.

Both the Food Chemicals Codex (FCC) and the Food and Agriculture Organization of the United Nations (FAO) have established specifications for the identity and purity of sodium carboxymethylcellulose for food use (Table 8).

B. Food Labeling

The accepted nomenclature for purified sodium carboxymethylcellulose that may be used on a food label ingredient statement in the U.S. by a manufacturer or processor is "sodium carboxymethylcellulose" or the shorter term "cellulose gum". These designations are permitted by both the FDA and FCC.

Establishment of "cellulose gum" as an accepted common name for sodium carboxymethylcellulose resulted from a Hercules petition granted by order of the Deputy Commissioner of Food and Drugs, effective June 26, 1963.

Although the terminology sodium carboxymethylcellulose, cellulose gum, and CMC are utilized interchangeably in this text, the name CMC is not advocated for use on a food label. This is so, because CMC also refers to industrial grades of the gum which do not meet the purity standards required for food use.

C. Standards for Food Use

Cellulose gum is utilized in a wide variety of food systems except in those items where a Standard of Identity precludes its use, e.g., pasta. Standards which include the use of cellulose gum are outlined below.

Title 9, Chapter III, of the Code of Federal Regulations lists ingredients acceptable for use in meat and poultry food products by the U.S. Department of Agriculture, subject to labeling requirements:

- Section 318.7 — Cellulose gum is included as a binder, extender, or stabilizer for meat and poultry baked pies when used in an amount sufficient for the purpose.
- Section 381.147 — Cellulose gum is included as a binder, extender, or stabilizer in various poultry products when used in an amount sufficient for the purpose.

Cellulose gum may be used in a wide variety of standardized foods subject to Title 21 of the Code of Federal Regulations. Within each of the following parts of "Subchapter B — Food and Food Products" are several food definitions and standards permitting use of cellulose gum:

- Part 131 — Milk and cream products
- Part 133 — Cheese and related cheese products
- Part 135 — Frozen desserts
- Part 169 — Food dressings and flavorings
- Part 150 — Fruit butters, jellies, and preserves

In addition, the following definitions and standards permit use of cellulose gum:

- Section 146.121 — Frozen concentrate for artificially sweetened lemonade
- Section 152.126 (A) — Frozen cherry pie
- Section 168.180 — Table syrup
- Section 165.175 — Soda water

Cellulose gum, including standard grades, is permitted for use in food packaging applications under the following regulations:

Table 8
FCC AND FAO SPECIFICATIONS FOR SODIUM CARBOXYMETHYLCELLULOSE

	FCC	FAO
Purity (%)	99.5	99.5
Degree of substitution	0.95(max)	0.20—1.00
Sodium, maximum (%)	9.5	—
Viscosity 2%, minimum (CPS)	25	—
Heavy metals, maximum (PPM)	40	40
Arsenic, maximum (PPM)	3	3
Lead, maximum (PPM)	10	10
Loss on drying, maximum (%)	10	12

- Section 174.5 — General provisions applicable to indirect food additives
- Section 175.105 — Adhesives
- Section 175.300 — Resinous and polymeric coatings
- Section 176.170 — Components of paper and paperboard in contact with aqueous and fatty foods
- Section 176.180 — Components of paper and paperboard in contact with dry food
- Section 177.1210 — Closures with sealing gaskets for food containers
- Section 182.70 — Substances migrating to food from cotton fabrics used in dry food packaging

D. Toxicological Aspects

Of all gums, sodium carboxymethylcellulose is one of the most extensively tested hydrocolloids for establishment of nontoxicity and safety in foods at the present time.

Subacute and chronic oral toxicity studies in rats, guinea pigs, dogs, and humans demonstrate that cellulose gum possesses no systemic toxic properties. Very high levels, such as 10% of the diet, possess a laxative effect consistent with the hydrophilic properties of the product and similar to other food gums. However, consumption of dietary amounts sufficient to produce a laxative effect would be completely unrealistic. Gastrointestinal absorption studies in rats and dogs demonstrate that cellulose gum is not absorbed from the intestinal tract. These studies lead to the conclusion that, on ingestion, cellulose gum is physiologically inert and noncaloric.

Various workers have established the toxicological background for purified sodium carboxymethylcellulose.[82-84] Toxicological investigations conducted on Hercules cellulose gum are summarized below.[47]

1. Skin Irritation and Sensitization

Two hundred human volunteers have been patch-tested by the Schwartz technique with no evidence of primary skin irritation or sensitization from cellulose gum.

2. Eye Irritation

Application into the eyes of New Zealand albino rabbits of 10 mg of cellulose gum in powder form produced slight irritation, which cleared completely in about 72 hr without washing out of the eyes.

Application into the eyes of 0.1 mg of cellulose gum, in the form of a 0.1% aqueous suspension, produced very slight irritation, which cleared completely in about 24 hr without washing out of the eyes.

3. Acute Oral Toxicity

Oral administration of suspensions of cellulose gum in olive oil were given to white rats and guinea pigs. The following results were noted:

Acute Oral LD$_{50}$

Rats	27 g/kg body weight
Guinea pigs	16 g/kg body weight

4. Six-Month Oral Toxicity

For 6 months, 100 rats, 100 guinea pigs, and 10 dogs were fed dietary levels of 1% and 2% (0.5 g and 1.0 g of cellulose gum/kg of body weight added to diet daily). Normal growth, fertility, urinalysis, and hematology were observed during the course of the experiment, and no gross or microscopic pathology was detected upon termination.

Three dogs were also fed dietary levels of 5 and 10% for 6 months. Pure grades of sodium alginate and karaya gum were fed for comparison at a level of 10%. At the 5% level, all observations during the experiment and at termination were normal. At the 10% level, growth was retarded with cellulose gum, as well as with sodium alginate and karaya gum. Loose stools were observed with all three products. No other changes (urinalysis, hematology, gross and microscopic pathology) were detected. Attempts to feed 20% cellulose gum in the diets of dogs were unsuccessful, owing to food refusal.

5. One-Year Oral Studies

For 1 year, 20 guinea pigs were fed dietary levels of 1 and 2% (0.5 g/kg daily added to the diet). No mortality occurred; growth was normal; and, upon termination, no gross or microscopic pathology was detected.

6. Chronic Oral Toxicity

For 2 years, 25 rats were fed dietary levels of 0.2, 1, and 2% (0.1 g/kg, 0.5 g/kg, and 1.0 g/kg added to their diet). Mortality, growth, monthly urinalysis, and hematology were normal. Gross and microscopic examination after 25 months' feeding revealed no pathology other than the senility changes present in controls. No neoplasms were found in any of the rats fed cellulose gum.

7. Reproduction

Rats fed dietary levels of 0.2, 1, and 2% cellulose gum were carried through three-generation reproduction studies, with offspring being maintained on these same dietary levels. No alterations in fertility or reproduction of these animals were detected.

8. Gastrointestinal Absorption

Radioactive (purified) cellulose gum, manufactured with ^{14}C-tagged sodium chloroacetate, was administered to rats as an aqueous solution at a dosage of 1.3 g/kg. Urine was collected in special metabolism cages to prevent cross-contamination with feces. The livers and kidneys were analyzed for radioactivity 44 hr after dosing. None was found at a sensitivity equivalent to less than 0.02% of the administered dose. Urine specimens showed an average activity equivalent to 0.14% of the administered dose. This activity corresponded to the amount of radioactive salt (sodium glycolate) formed in the synthesis, and was established by chromatography of the urine solid not to be NaCMC or other saccharide polymer.

9. Clinical Studies

For 6 months, 11 human volunteers ingested 10 g of cellulose gum daily. Complete hematology and urinalysis on all subjects revealed no alterations during administration. Bone marrow studies were made on three subjects during the test and six subjects at completion; all were within normal limits. Some additional volunteers experienced a laxative effect at 10 g daily. No other physiological manifestations were detected.

Sodium carboxymethylcellulose in powder form was used as a vehicle for an antiseptic substance in 134 cases treated for vaginal infections of several types. Five grams of the material was used for each subject. There was no evidence of irritation or untoward reactions to the vaginal mucosa or external genitalia of any of the cases.

10. Toxicity to Fish

Employing a standard laboratory procedure, rainbow trout and bluegills were exposed under static conditions to technical-grade CMC at several concentrations, up to a maximum level of 100 ppm, and observed for mortality or other adverse reaction over a period of 96 hr. Results of this 96-hr observation demonstrate that CMC has a 4-day median tolerance limit (TL_{50}) of greater than 100 ppm, the highest level tested. In addition, no adverse reactions were noted in the fish exposed to CMC. These results demonstrate that CMC has a low order of toxicity to fish.

REFERENCES

1. **Sundgerg, B.,** *Farm. Revy,* 45, 8247, 1946.
2. **Jansen, E.,** Deutsche Celluloid Fabrik Eilenberg, German Patent 332,203, 1918.
3. **Dicker, R.,** *Rayonne,* 6(2), 73; 6(3), 77, 1950.
4. **Bartholome, E. and Buschman, K. F.,** *Melliand Textilber.,* 30, 249, 1949.
5. **Burt, L. H.,** *Drug Stand.,* 19, 106, 1951.
6. **Stelzer, G. I. and Klug, E. D.,** Sodium carboxymethylcellulose, in *Handbook of Water Soluble Gums and Resins,* Davidson, R., Ed., McGraw-Hill, New York, 1980, chap. 4.
7. United States Tariff Commission Reports 1968, U.S. Government Printing Office, Washington, D.C., 1968.
8. *Chemical Economics Handbook,* Stanford Research Institute, Menlo Park, Calif., 1981.
9. **Whistler, R. L.,** *Industrial Gums,* Academic Press, New York, 1959.
10. **Glicksman, M.,** Cellulose gums, in *Gum Technology in the Food Industry,* Academic Press, New York, 1969, chap. 12.
11. **Manners, D. J.,** Cellulose and glucans, in *Recent Advances in Food Science,* Vol. 3, Leitch, J. M. and Rhodes, D. N., Eds., Butterworth, London, 1963, 291—299.
12. **Baird, G. S. and Speicher, J. K.,** Carboxymethylcellulose, in *Water Soluble Resins,* Davidson, R. L. and Sittig, M., Eds., 1962.
13. **Whistler, R. L. and Smart, C. L.,** *Polysaccharide Chemistry,* Academic Press, New York, 1953.
14. **Reuben, J. and Connor, H. T.,** Analysis of the carbon-13 NMR spectrum of hydrolyzed 0-(carboxymethyl) cellulose: monomer composition and substitution patterns, *Carbohydr. Res.,* 115, 1, 1983.
15. **Ganz, A. J.,** U.S. Patent 3,485,651, 1969.
16. **Whelan, K.,** U.S. Patent 3,503,895, 1970.
17. **Hoefler, A.,** unpublished data, Hercules Inc., 1982.
18. **Ganz, A. J.,** Cellulose gum — a texture modifier, *Manuf. Confect.,* 46(10), 23—33, 1966.
19. **Ganz, A. J.,** Some effects of gums derived from cellulose on the texture of foods, *Cereal Sci. Today,* 18(12), 398, 1973.
20. Hercules Inc., Cellulose Gum — Chemical and Physical Properties, 1981.
21. Hercules Inc., Technical Data Bull. No. FF 321, 1981.
22. **Elliot, J. H. and Ganz, A. J.,** Modification of food characters with cellulose hydrocolloids. I. Rheological characterization of an organoleptic property, *J. Texture Stud.,* 2, 220, 1971.
23. **Elliot, J. H. and Ganz, A. J.,** Some rheological properties of sodium carboxymethylcellulose solutions and gels, *Rheol. Acta,* 13, 1178, 1974.
24. **Elfak, A. M., Pass, G., and Phillips, G. O.,** The effect of shear rate on the viscosity of sodium carboxymethylcellulose and κ-carrageenan, *J. Sci. Food Agric.,* 30, 724, 1979.
25. **Igoe, R. S.,** Hydrocolloid interactions useful in food systems, *Food Technol.,* 36(4), 72, 1982.
26. **Elliot, E. S.,** U.S. Patent 2,639,239, 1953.

27. **Shenkenbertg, D. R., Chang, J. C., and Edmondson, L. F.,** Develops Milk-Orange Juice, U.S. Patent 3,692,532, 1972.
28. **Matz, S. A.,** *Food Texture,* AVI, Westport, Conn., 1962.
29. **DeButts, E. H., Hudy, J. A., and Elliot, J. H.,** *Ind. Eng. Chem.,* 49, 94, 1957.
30. **Nijhof, G. J. J.,** U.S. Patent 3,418,133, 1968.
31. Hercules Inc., Technical Data Bull. No. FF 306A, 1981.
32. **Imeson, A. P., Ledward, D. A., and Mitchell, J. R.,** On the nature of the interaction between some anionic polysacharides and proteins, *J. Sci. Food Agric.,* 28(8), 661, 1977.
33. **Ganz, A. J.,** U.S. patent 3,407,076, 1968.
34. **Thompson, R. E.,** U.S. Patent 2,824,092, 1958.
35. **Hill, R. D. and Zadow, J. G.,** The recovery of proteins from cheddar cheese whey by complex formation with carboxymethylcellulose, *Aust. J. Dairy Technol.,* 33(3), 97, 1978.
36. **Zadow, J. G. and Hill, R. D.,** The formation of complexes between whey proteins and carboxymethyl-cellulose modified with substitutents of increased hydrophobicity, *J. Dairy Res.,* 45(1), 85, 1978.
37. **Craver, L. A.,** Canadian Patent 995971.
38. **Hansen, P. M. T., Hidalgo, J., and Gould, I. A.,** Reclamation of whey protein with carboxymethyl-cellulose, *J. Dairy Sci.,* 54(6), 830, 1971.
39. **Hidalgo, J. and Hansen, P. M. T.,** *J. Agric. Food Chem.,* 17, 1089, 1969.
40. **Zadow, J. G. and Hill, R. D.,** Complex formation between whey proteins and insoluble cross linked carboxymethylcellulose, *N.Z. J. Dairy Sci. Technol.,* 13, 162, 1978.
41. **Asano, Y.,** The interaction between milk proteins and carboxymethylcellulose in fruit-flavored milk, *Proceedings* 17th *International Dairy Congress,* pp 695-702, Munich, (1960).
42. **Pedersen, J. K.,** Process for Preparing Gelled Sour Milk, U.S. Patent 3,918,243, 1976.
43. **Ganz, A. J.,** U.S. Patent 3,287,139, 1966.
44. **Desmarais, A. J. and Ganz, A. J.,** Effect of cellulose gum on sugar crystallization and its utility in confections, *Manuf. Confect.,* 46(10), 23, 1962.
45. **Keller, J.,** unpublished data, Hercules Inc., 1981.
46. **Ganz, A. J.,** CMC and hydroxypropylcellulose — versatile gums for food use, *Food Prod. Dev.,* 3(6), 65, 1969.
47. Hercules Inc., Technical Data Bull. No. FF 342, 1981.
48. **Podlas, T. J.,** Process of Preparing CMC Gels, U.S. Patent 3,719,503, 1973.
49. **Sleap, J.,** personal communication, Hercules Inc., 1983.
50. **Arbuckle, W. S.,** *Ice Cream,* AVI, Westport, Conn., 1977.
51. **Frandsen, J. H. and Markham, S. A.,** *The Manufacture of Ice Cream and Ices,* Orange Judd, New York, 1915.
52. **Potter, F. E. and Williams, D. H.,** Stabilizers and emulsifiers in ice cream, *Milk Plant Mon.,* 39(4), 76, 1950.
53. **Moncrieff, R. W.,** Ice cream stabilizers, *Food Manuf.,* 29(8), 314, 1954.
54. **Sperry, G. O.,** Stabilizers and HTST, *Ice Cream Field,* 65(5), 10, 1955.
55. **Moss, J. R.,** Stabilizers and ice cream quality, *Ice Cream Trade J.,* 51, 22, 1955.
56. **Boyle, J. L.,** The stabilization of ice cream and ice lollies, *Food Technol. Aust.,* 11, 543, 1959.
57. **Frandsen, J. H. and Arbuckle, W. S.,** *Ice Cream and Related Products,* AVI, Westport, Conn., 1961.
58. **Cottrell, J. I., Pass, G., and Phillips, G. O.,** The effect of stabilizers on the viscosity of an ice cream mix, *J. Sci. Food Agric.,* 31(10), 1066, 1980.
59. **Moore, L. J. and Shoemaker, C. F.,** Sensory textural properties of stabilized ice cream, *J. Food Sci.,* 46(2), 399, 1981.
60. **Doane, F. J. and Keeney, P. G.,** Frozen dairy products, in *Fundamentals of Dairy Chemistry,* Webb, B. H. and Johnson, A. H., Eds., AVI, Westport, Conn., 1965.
61. **Shipe, W. F., Roberts, W. M., and Blanton, L. F.,** Effect of ice cream stabilizers on the freezing characteristics of various aqueous systems, *J. Dairy Sci.,* 46(3), 169, 1963.
62. **Nickerson, T. A.,** Lactose crystallization in ice cream. IV. Factors responsible for reduced incidence of sandiness, *J. Dairy Sci.,* 45(3), 354, 1962.
63. **Cottrell, J. I., Pass, G., and Phillips, G. O.,** Assessment of polysaccharides as ice cream stabilizers, *J. Sci. Food Agric.,* 30, 1085, 1979.
64. **Josephson, D. V. and Dahle, C. D.,** A new cellulose gum stabilizer for ice cream, *Ice Cream Rev.,* 28(11), 32, 1945.
65. **Pompa, A.,** Recipes for ice cream using sodium carboxymethylcellulose as the stabilizer, *Food,* 14, 231, 1945.
66. **Burt, L. H.,** Frozen Confections, U.S. Patent 2,548,865, 1951.
67. **Werbin, S. J.,** Recent advances in the field of stabilizers and emulsifiers, *South. Dairy Prod. J.,* 53(3), 38, 1953.
68. **Glicksman, M.,** Utilization of synthetic gums in the food industry, *Adv. Food Res.,* 12, 283, 1963.

69. **Dahle, C. D.,** Analysis of stabilizers, *Ice Cream Field.,* 48(4), 62, 1946.

70. **Rothwell, J. and Palmer, M. M.,** Modern trends in ice cream stabilizers, *Dairy Ind.,* 30(2), 107, 1965.

71. **Blihovde, N.,** Process for Stabilizing Foodstuff and Stabilizing Composition, U.S. Patent 2,604,406, 1952.

72. **Keeney, P. G.,** Effect of some citrate and phosphate salts on stability of fat emulsions in ice cream, *J. Dairy Sci.,* 45(3), 430, 1962.

73. **Finney, D. J.,** U.S. Patent 3,993,793, 1976.

74. **Landers, M.,** Readily Water Soluble Carboxymethylcellulose, U.S. Patent 2,445,226, 1948.

75. **Steinitz, W. S.,** Stabilizers for Ice Cream and Sherbet, U.S. Patent 2,823,129, 1958.

76. **Baugher, W. L.,** Low Density Frozen Dessert, U.S. Patent 3,968,266, 1976.

77. **Bundus, R. H.,** Frozen Dessert Composition, U.S. Patent 4,307,123, 1981.

78. **Anon.,** Corn-soy snacks and peanut-based products, *Food Process. (Chicago),* 39(1), 58, 1978.

79. **Boyle, J. L.,** The stabilization of ice cream and ice lollies, *Food Technol. Aust.,* 11, 543, 1959.

80. **Klose, R. E. and Glicksman, M.,** Gums, in *Handbook of Food Additives,* Chemical Rubber Co., Cleveland, Ohio, 1968.

81. **Burt, L. H.,** Frozen Confections, Canadian Patent 527,490, 1956.

82. **Shelanski, H. A. and Clark, A. M.,** Physiological action of sodium carboxymethylcellulose on laboratory animals and humans, *Food Res.,* 13(1), 29, 1948.

83. **Brown, C. J. and Houghton, A. A.,** The chemical and physical properties of carboxymethylcellulose and its salts, *J. Soc. Chem. Ind.,* 60, 254, 1941.

84. **Rowe, V. K., Spencer, H. C., Adams, E. M., and Irish, D. D.,** Response of laboratory animals to cellulose glycolic acid and its sodium and aluminum salts, *Food Res.,* 9(3), 175, 1944.

85. **Braverman, A.,** Flavored Freezable Gel Confection, U.S. Patent 4,140,807, 1979.

86. Hercules Inc., Guide to Food Uses for Hercules Cellulose Gum, 1983.

87. **Marulich, A. J.,** U.S. Patent 3,826,829, 1974.

88. **Howler, B., Wasserman, G. S., and Keehner, J. E.,** U.S. Patent 3,897,571, 1975.

89. **Rubenstein, I. H.,** U.S. Patent 3,525,624, 1970.

90. **LeVan, D. J., Homler, B. E., and Bosco, P. M.,** U.S. Patent 3,619,205, 1971.

91. **Keller, J.,** unpublished data, Hercules Inc., 1980.

92. Hercules Inc., Technical Data Bull. No. FF 312, 1981.

93. **McCluskey, F. J., Thomas, E. L., and Coulter, S. T.,** Precipitation of milk proteins by sodium carboxymethylcellulose, *J. Dairy Sci.,* 52, 1181, 1969.

94. **Smith, W. B.,** U.S. Patent 3,385,714, 1968.

95. **Stewart, A. P., Dreier, C. R., and Falk, J. D.,** U.S. Patent 3,666,497, 1972.

96. Cooperative Fabrick van Melkproducten Te Bedum, Preparing a Seed Material for Lactose Crystallization in Concentrated Milk Products, Dutch Patent 75,711 *Chem. Abstr.,* 49, 6504a, 1954.

97. **Asano, Y.,** Interaction between casein and carboxymethylcellulose in the acidic condition, *Agric. Biol. Chem.,* 34(1), 102, 1970.

98. **Morgan, D. R., Anderson, D. L., and Mook, D. E.,** Preparation of Acidified Milk Products, U.S. Patent 3,539,363, 1970.

99. **Cajigas, S. D.,** Instant Yogurt Drink Composition, U.S. Patent 4,289,789, 1981.

100. **Nutricia, N. V.,** Acid Milk Beverage, Dutch Patent 99,203 *(Chem. Abstr,* 56,1816b), 1962.

101. **Abdel Baky, A. A., El Fak, A. M., Abo El-Ela, W. M., and Farag, A. A.,** Fortification of Domiati cheese milk with whey proteins/carboxymethylcellulose complex, *Dairy Ind. Int.,* 49(9), 29, 1981.

102. **Pyler, E. J.,** Cake ingredients, in *Baking Science and Technology,* Vol. 2, Sibel, Chicago, 1973, chap. 22.

103. **Jaeger, E. B.,** Dry Mix, U.S. Patent 2,611,704, 1952.

104. **Bayfield, E. G.,** Improving white layer cake quality by adding CMC (carboxymethylcellulose), *Baker's Dig.,* 36(2), 50, 1962.

105. **Young, W. E. and Bayfield, E. G.,** Hydrophilic colloids as additives in white layer cakes, *Cereal Chem.,* 40, 195, 1963.

106. **Dubois, D. K.,** *Baker's Dig.,* 40(5), 73, 1966.

107. **Elsesser, C. C.,** Cake Mix and Method of Preparing Same, U.S. Patent 2,996,384, 1961.

108. **Hager, R. E. and Lowrey, E. R.,** U.S. Patent 3,071,472, 1963.

109. **Trimbo, H. B., Ma, S., and Miller, B. S.,** *Baker's Dig.,* 40(1), 40, 1966.

110. **Miller, B. S., Trimbo, H. B., and Sandstedt, R. M.,** The development of gummy layers in cakes, *Food Technol.,* 21, 377, 1967.

111. **Goodman, A. H., Pettinate, E. F., Kroll, A., and Lipka, D. H.,** U.S. Patent 3,378,378, 1968.

112. **Gass, R. L.,** U.S. Patent 3,343,965, 1967.

113. **Weigle, D. C.,** U.S. Patent 3,794,741, 1974.

114. **Dubois, D. K.,** Icings and glazes: formulation and processing, *Cereal Foods World,* 25(7), 390, 1980.

115. **Dubois, D. K.,** Icings and glazes for sweet yeast raised bakery foods. I. Ingredient functions, Tech. Bull., American Institute of Baking, 2(5), 1980.

116. **Birnbaum, H.,** The functional characteristics of icing stabilizers, *Baker's Dig.,* 34(2), 62, 1969.

117. **Lipman, H. J.,** Advances in icing technology, *Baker's Dig.,* 46(1), 47, 1972.

118. **Wagner, W. W.,** Boiled Icing and Method of Making the Same, U.S. Patent 2,682,472, 1954.

119. **Butler, R. W.,** Stabilized Icings and Preparations, U.S. Patent 2,914,410, 1959.

120. **Ganz, A. J.,** Stabilized Icings and Process, U.S. Patent 3,009,812, 1961.

121. **Grossi, F. X.,** Color Stabilization in Food Products, U.S. Patent 2,841,449, 1958.

122. **Krueger, J.,** Food grade gums, *Process. Prep. Foods,* 148(8), 66, 1979.

123. **Wheeler, F. G. and Endres, J. G.,** Icings, a new look at some basic aspects, *Baker's Dig.,* 39(5), 52, 1965.

124. **Morley, R. G.,** Troubleshooting icings and glazes, *Food Eng.,* 53(5), 78, 1981.

125. Powder for Meringue Whip and Marshmallow, Hercules Food Gums Recipe Guide, Bulletin FRG-115, 1980.

126. **Moyls, A. W., Atkinson, F. E., Strachan, C. C., and Britton, D.,** Preparation and storage of canned berry and berry-apple pie fillings, *Food Technol.,* 9, 269, 1955.

127. **Bisno, L.,** *Baker's Dig.,* 34(4), 44(5), 70, 1960.

128. **Kunz, C. E. and Robinson, W. B.,** Hydrophilic colloids in fruit pie fillings, *Food Technol.,* 16(7), 100, 1962.

129. **Rigler, L. E., Taki, G. H., and Spirtos, N. G.,** U.S. Patent 3,928,252, 1975.

130. **Schuurink, F. A.,** Improving Flour or Dough, Dutch Patent 59,870 *(Chem. Abstr.* 41:7014g), 1947.

131. **Bayfield, E. G.,** Gums and some hydrophilic colloids as bread additives, *Baker's Dig.,* 32(6), 42, 1958.

132. **VanHaften, J. L.,** Fat based food emulsifiers, *J. Am. Oil Chem. Soc.,* 56, 831A, 1979.

133. **vonRymon Lipinski, G.,** Process of Preparing a Baker Flour-Containing Product, U.S. Patent 4,172,154, 1979.

134. **Rubio, M. J.,** Tortilla and Process Using Edible Hydrophilic Gum, U.S. Patent 3,655,385, 1972.

135. **Tims, O. L.,** Flour Tortillas, U.S. Patent 4,241,106, 1980.

136. **Suderman, D. R., Wiker, J., and Cunningham, F. E.,** Factors affecting adhesion of coating to poultry skin: effects of various protein and gum sources in the coating composition, *J. Food Sci.,* 46, 1010, 1981.

137. **Dubois, D. K.,** Bread and Additive, U.S. Patent 3,219,455, 1965.

138. **Chang, P. K.,** Replacement of Egg Albumen in Food Compositions, U.S. Patent 4,214,009, 1980.

139. **Collins, J. T. and Hyldon, R. G.,** Instant Grits, U.S. Patent 3,526,512, 1970.

140. **Huffman, G. W. and Moore, J. W.,** Instant Oatmeal, U.S. Patent 2,999,018, 1961.

141. **Roundy, Z. D. and Osmond, N. R. H.,** Cheese Products, U.S. Patent 2,956,885, 1960.

142. **Daehler, R. A.,** New milk-based drink uses high fructose corn syrup, *Dairy Ice Cream Field,* 162(4), 64, 1979.

143. **Nishiyama, K.,** Apple Juice Composition and Milk-Apple Juice Drink Containing Such Compositions, U.S. Patent 4,078,092, 1978.

144. **Campbell, A. E. and Ford, M. A.,** British Patent 1,491,287, 1977.

145. **Common, J. L.,** Powdered Instant Beverage Mix, U.S. Patent 3.023,106, 1962.

146. **Jackman, K. R.,** Citrus Fruit Juice and Drink, U.S. Patent 4,163,807, 1979.

147. **Sirett, R. R., Eskritt, E. J., and Derlatka, E. J.,** Dry Beverage Mix Composition and Process, U.S. Patent 4,264,638, 1981.

148. **Horikoshi, K.,** An Edible Emulsion of Perfume Oil, Japanese Patent 4179 *(Chem. Abstr.* 49, 10592a), 1954.

149. **Stewart, A. P. and Falk, J. D.,** U.S. Patent 3,669,675, 1972.

150. **Wallerstein, J. S., Schade, A. L., and Levy, H. B.,** Process for Improving the Foam of Fermented Beverages, U.S. Patent 2,547,988, 1951.

151. **Pangborn, R. M., Gibbs, Z. M., and Tassan, C.,** Effect of hydrocolloids on apparent viscosity and sensory properties of selected beverages, *J. Texture Stud.,* 9, 415, 1978.

152. **Glicksman, M. and Farkus, E.,** Gums in artificially sweetened foods, *Food Technol.,* 20, 58, 1966.

153. **Brenner, G. M.,** Sugarless Beverage, U.S. Patent 2,691,591, 1954.

154. **Bernotavicz, J. W.,** Shaped Blood By-Product and Process, U.S. Patent 4,143,168, 1979.

155. **Bone, D. P. and Shannon, E. L.,** Process for Making a Dry Pet Food Having a Hard Component and a Soft Component, U.S. Patent 4,006,266, 1977.

156. **Burkwall, M. P.,** Simulated Egg and Meat Pet Food, U.S. Patent 3,974,296, 1976.

157. Hercules Inc., Technical Data Bull. No. FFD-202a, 1982.

158. **Dannelly, C. C.,** Rumen-Stable Pellets, U.S. Patent 4,181,709, 1980.

159. **Anderson, E. E., Blank, A. P., and Esselen, W. B.,** Quell separation in pickle relish, *Food Eng.,* 26(4), 131, 1954.

160. Hercules Inc., Technical Data Bull. No. FF 308c, 1982.

161. **Andersen, G.,** Sugar-Free Confectionery Material Based on Xylitol and Sorbitol, U.S. Patent 4,292,337, 1981.

162. **Ferguson, E. A.,** Appetite Satient, U.S. Patent 2,714,083, 1955.

163. **Rispoli, J. M., Sabhlok, J. P., Ho, A. S., Scherer, B. G., and Guiliano, C.,** Oil Replacement Composition, U.S. Patent 4,308,294, 1981.

164. **Akata, S. and Koda, R.,** Utilization of strawberries. II. Methods of preserving the organoleptic qualities and ascorbic acid content of frozen strawberries, *J. Ferment. Technol. Japan,* 32, 132 (*Chem. Abstr.* 46:7250c), 1954.

165. **Burt, L. H.,** Processed Food and Method of Preparation, U.S. Patent 2,728,676, 1955.

166. **Latham, S. D. and Seeley, R. D.,** U.S. Patent 3,769,404, 1973.

167. **Seeley, R. D. and Seeley, R. B.,** Cholesterol-Free Egg Product Having Improved Cooking Tolerance, U.S. Patent 4,200,463, 1980.

168. **Shemwell, G. A. and Stadelman, W. I.,** Some factors affecting weepage of baked spinach souffles, *Poultry Sci.,* 55(4), 1467, 1976.

169. **Perech, R.,** Vegetable or Fruit Juice Concentrate, U.S. Patent 2,393,561, 1946.

170. **Bogin, H. H. and Feick, R. D.,** Fatty Food Composition and Method of Making the Same, U.S. Patent 2,555,467, 1951.

171. **Galle, E. L., Mikkelson, M. O. and Kolasky, J. F.,** U.S. Patent 3,719,497, 1973.

172. **Algemene Kunstzijde Unie N.V.,** Preserving Fish and Meat in Cans, Dutch Patent 69,185 *Chem. Abstr,* 46:7250c, 1951.

173. **Kojima, Y., Inamasu, Y., and Shiraishi, T.,** Agar gel in canned mitsumane. I. Agents reinforcing agar gel., *Norinsho Suisan Koshusho Kinkyu Hokoku* 8,161 (*Chem. Abstr.* 55:8691d), 1951.

174. **Gejl-Hansen, F. and Flink, J. M.,** Microstructure of freeze dried emulsions: effect of emulsion composition, *J. Food Process. Preserv.,* 2(3), 205, 1978.

175. **Bomstein, R. A.,** Texturizing Materials for Foods, U.S. Patent 3,833,744, 1974.

176. **Pader, M., Hamilton, H. D., Miles, J. J., and McCrimlisk, G. J.,** U.S. Patent 2,874,049, 1959.

177. **Mason, D. F.,** Fruit Preservation Process, U.S. Patent 3,472,662, 1969.

178. **Cornwell, R. T. K.,** Protective Coating Composition for Hams, U.S. Patent 2,558,012, 1951.

179. **Algemene Kunstzij de Unie N.V.,** Preserving Eggs, Dutch Patent 70,064 (*Chem. Abstr.* 46:10485), 1952.

180. **Vale, W. H.,** Coating Composition for Fruits and Vegetables, Australian Patent 153,174, 1953.

181. **Shinn, B. M. and Childs, W. H.,** Meat Coating Composition and Method, U.S. Patent 2,721,142, 1955.

182. **Brown, C. F., Gooding, C. and Vahlteich, H. W.,** Sorbic Acid Antimold Solution, U.S. Patent 2,856,294, 1958.

183. **Nikkila, O. W., Linko, R. R., and Jaurola, V. E.,** Treatment of Fresh Food which is to be Frozen, Finnish Patent 31,427, 1960.

184. **Ghosh, K. G., Srivatsa, A. N., Nirmala, N., and Sharma, T. R.,** Development and application of fungistatic wrappers in food preservation. II. Wrappers made by coating process, *J. Food Sci. Techol. India,* 14(6), 261, 1977.

185. **Hutchison, B. R. and Swanson, A. M.,** Coated Fibrous Cellulose Product and Process, U.S. Patent 4,143,163, 1979.

186. Charles Pfizer & Co., Inc., Limmits, A Satisfying Food for Losing Weight, Family Products Division, New York, 1962.

187. **Batdorf, J. B.,** How cellulose gum can work for you, *Food Eng.,* 36(8), 66, 1964.

188. **Katayama, S., Matsukawa, H., Yamamoto, M., and Matsuyama, J.,** U.S. Patent 3,687,865, 1972.

189. **Hiestand, E. N., Jensen, E. H., and Meister, P. D.,** U.S. Patent 3,549,555, 1970.

190. **Ohtaki, S.,** Process for Manufacturing Powdered Preparations Containing Fat-Soluble Vitamins, Essential Oils and Mixtures Thereof, U.S. Patent 3,056,728, 1962.

191. **Palmer, E.,** Method for Encapsulating Materials, U.S. Patent 3,989,852, 1976.

192. **Boyd, J. W., Bissett, H. M., and Tarr, H. L. A.,** Further observations on the distribution of chlortetracycline throughout ice blocks, *Fish. Res. Board Can. Prog. Rep. Pac. Coast Stn.,* 102, 14, 1955.

193. **Aktieselskabet Protan,** Preserving Food Products by Freezing in Thickened Salt Solution, Norwegian Patent 86,681 (*Chem. Abstr.* 50:7348a), 1955.

194. **Spradling, A. B.,** Coating Material for Tablets, U.S. Patent 2,693,437, 1954.

195. **Spradling, A. B.,** Coating Material for Tablets, U.S. Patent 2.693,436, 1954.

196. **Doerr, D. W., Serles, E. R., and Deardorff, D. L.,** Tablet coatings: cellulosic high polymers, *J. Am. Pharm. Assoc.,* 43(7), 434, 1954.

197. **Long, S.,** Tablets Coated with Carboxymethylcellulose Shellac Compositions, U.S. Patent 3.043,747, 1962.

198. **Christenson, G. L. and Dale, L. B.,** Sustained Release Tablet, U.S. Patent 3,065,143, 1962.

199. **Jonas, J. J.,** U.S. Patent 4,012,533, 1977.

200. **Strums, R. L. and Jonas, J. J.,** U.S. Patent 3,991,224, 1976.

201. **Gilmore, C. and Miller, D. E.,** Whippable Topping Mix, U.S. Patent 4,208,444, 1980.

202. **Pader, M. and Gershon, S. D.,** U.S. Patent 3,224,883, 1965.
203. **Igoe, R. S.,** Acidified Milk Gel and Method of Producing the Same, U.S. Patent 3,996,390, 1976.
204. **Igoe, R. S.,** Acidified Milk Product and Method of Producing the Same, U.S. Patent 4,046,925, 1977.
205. **Trudsoe, J. E.,** unpublished data, Copenhagen Pectin Factory, Lille Skensved, Denmark, 1983.
206. **Glicksman, M.,** Freezable Gels, U.S. Patent 3,060,032, 1962.
207. **Sleap, J.,** personal communication, Hercules Inc., 1984.
208. **Matsumoto, K. and Matsuura, K.,** Fortified Artificial Rice, Japanese Patent 4170 (*Chem. Abstr.* 49:1054g), 1954.
209. **Taninaka, A. and Hori, M.,** Rice Substitute, Japanese Patent 9230 (*Chem. Abstr.* 55:844b), 1958.
210. **Battista, O. A. and Smith, P. A.,** Level-Off DP Cellulose Products, U.S. Patent 2,978,446, 1961.
211. **Battista, O. A. and Smith, P. A.,** Microcrystalline cellulose, *Ind. Eng. Chem.,* 54(9), 20, 1962.
212. **McGinley, E. J. and Zuban, J. M.,** Method of Mixing Difficultly Dispersable Material, e.g., Fat or Wax, in a Gum-Microcrystalline Cellulose Matrix, and Powder Product, U.S. Patent 4,231,802, 1980.
213. **Bishop, T. R.,** New Zealand Patent 151,719, 1971.
214. **Leipold, D. P.,** personal communication, Hercules Inc., 1984.

Chapter 3

HYDROXYPROPYLCELLULOSE (HPC)

Christopher McIntyre

TABLE OF CONTENTS

I. BACKGROUND

The preceding chapter dealt with the properties and functionality of sodium carboxyme-thylcellulose, an ionic polymer. The subject of this chapter is hydroxypropylcellulose, a nonionic hydrocolloid which has unique functional properties that permit its use in applications different from those of other cellulosic polymers used in foods.

II. DESCRIPTION

Hydroxypropylcellulose (trademark Klucel®, Hercules, Inc.) is a nonionic water-soluble cellulose ether with an interesting combination of properties. It combines organic solvent solubility, thermoplasticity, and surface activity with the thickening and stabilizing properties characteristic of other water-soluble cellulose polymers.

III. REGULATORY STATUS

Hydroxypropylcellulose designated for use in food applications conforms to the specifications set forth in U.S. Code of Federal Regulations Title 21, Section 172-870. It may be used as a food additive to serve several functions. It generally may be used in foods except for certain standardized foods whose regulatory definitions do not currently provide for use of this gum. The Food Chemical Codex also lists specifications for food grade hydroxypropylcellulose.

IV. MANUFACTURE

The manufacture of hydroxypropylcellulose is carried out by reacting propylene oxide with alkali-treated cellulose under conditions of elevated pressure and temperature such that the propylene oxide is substituted at the reactive hydroxyl sites on the anhydroglucose monomer units. There are three hydroxyl sites on each anhydroglucose unit. Thus a degree of substitution (DS) of three can be achieved. The hydroxypropyl substituent groups contain secondary hydroxyls which are available for further reaction with propylene oxide, leading to the formation of side chains that contain more than one hydroxypropyl group. Thus hydroxypropylcellulose can have a molar substitution (MS) of more than 3.

V. STRUCTURE

As mentioned above, the reactive hydroxyls on the anhydroglucose units of cellulose may become substituted. The remaining hydroxyl groups on the polymer are secondary. This is seen in Figure 1 in an idealized structure with a molar substitution of three. The molecular weight of hydroxypropylcellulose can vary from 60,000 to approximately 1,000,000.

VI. PROPERTIES

A. Solubility

Hydroxypropylcellulose is soluble in water below 38°C. Solutions are clear, smooth and nonthixotropic. It is insoluble in hot water and precipitates as a swollen floc at temperatures of between 40 and 45°C.

Like other hydrocolloids, hydroxypropylcellulose has a tendency to agglomerate when the dry powder is added to water. Hydration of the outer surfaces can cause lumping. Hydroxypropylcellulose is soluble in water at room temperature, but is insoluble in water above 45°C. This property is used in one method of preparing solutions as follows:

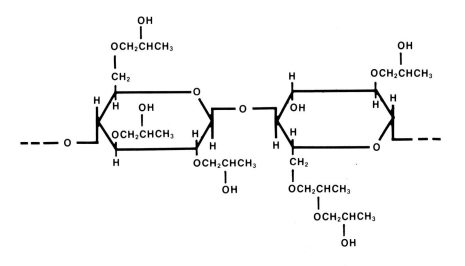

FIGURE 1. Idealized structure of hydroxypropylcellulose (MS 3.0).

- Method 1. The powder is slurried in hot water before adding to the main volume of water. Using this technique, a high solids slurry can be produced by adding dry hydroxypropylcellulose powder to about six times its own weight of hot water at a temperature of 50 to 60°C. The slurry should be well agitated and stirred for a few minutes before adding to the required volume of cold water to form the final solution. Slurrying in a nonsolvent disaggregates the particles which results in a faster dissolution when the cold water is added. The temperature during slurrying should be maintained above 50°C to insure that no partial dissolution of particles occurs resulting in the formation of gels or fisheyes. The hot slurry is then diluted with cold water at room temperature or lower. Agitation is continued until all particles are fully dissolved and the solution is free of gels. Hydroxypropylcellulose is surface active and, therefore, solutions have a tendency to foam. For this reason, high-shear agitation should be avoided. It is more effective to agitate long enough for full dissolution to occur than to use a high-shear stirrer. The time required for complete dissolution depends on solution concentration and the viscosity range of the gum being used. Lower viscosity types at low concentrations require the shortest time for complete dissolution.

- Method 2. Hydroxypropylcellulose can be added to the vortex of well-agitated water at room temperature. The rate of addition must be slow enough to permit individual particles to become separated in the water. Addition of the powder should, however, be completed before any appreciable viscosity buildup is obtained in the solution, otherwise separation of particles becomes more difficult and gels may be formed. When all the powder is added, the rate of agitation should be reduced, but stirring should be continued until a gel-free solution is obtained. Throughout the mixing period, solution temperature should be maintained below 35°C.

- Method 3. Dry blend hydroxypropylcellulose with any inert or nonpolymeric soluble material that will be used in the formulation. Blending aids the separation of the gum particles during the first wetting stage and thus reduces the chances of lump formation. An effective ratio is 1 part of hydroxypropylcellulose to 4 or 5 parts of diluent. The blend is dispersed into solution as described in Method 2.

B. Viscosity

Hydroxyropylcellulose has excellent solubility in water and in many polar organic solvents. Solutions are clear, smooth flowing, nonthixotropic and free from gels and fibers. Solutions

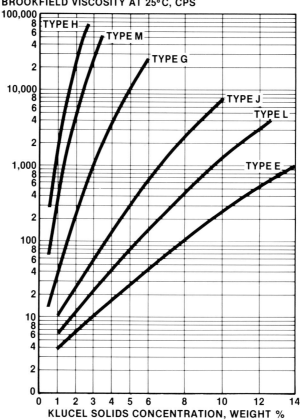

FIGURE 2.   Effect of concentration and Klucel® type of viscosity of water solutions.

are non-Newtonian in flow, and show changes in apparent viscosity with changes in shear rate. Solutions can be prepared with a wide range of viscosities depending on the concentration and the viscosity grade used.

1. Effect of Concentration

Viscosity of solutions increases rapidly with concentration and becomes an almost straight-line relationship when plotted on a semilog basis. This is shown in Figure 2.

2. Temperature (Precipitation)

As with most polymers, the viscosity of aqueous solutions decreases as temperature is increased. However, at temperatures in the range of 40 to 45°C, hydroxypropylcellulose is precipitated from solution. This precipitation is completely reversible. The hydrocolloid redissolves when the solution is cooled below 40°C with stirring, and the original viscosity is obtained. Precipitation first appears as cloudiness in the aqueous solution accompanied by a marked reduction in viscosity. The polymer does not gel when it precipitates from solution. The low-viscosity types separate as a highly swollen and finely divided precipitate. High-viscosity types, particularly if agitation is present, may form a stringy precipitate. The addition of cellulose gum reduces the tendency for agglomeration of the gum as it precipitates. Using this effect, the high-viscosity types can also be precipitated in a finely divided form. The presence of relatively high concentrations of other dissolved materials, such as salt or sucrose, results in a lowering of the precipitation temperature. The magnitude of the lowering

Table 1
COMPARATIVE EFFECT OF SOLUTION
COMPOSITION ON PRECIPITATION
TEMPERATURE

Ingredients and concentration	Precipitation temperature, °C
1% Klucel H®	41
1% Klucel H® + 1.0% NaCl	38
1% Klucel H® + 5.0% NaCl	30
0.5% Klucel H® + 10% sucrose	41
0.5% Klucel H® + 20% sucrose	36
0.5% Klucel H® + 30% sucrose	32
0.5% Klucel H® + 40% sucrose	20
0.5% Klucel H® + 50% sucrose	7

Courtesy of Hercules, Inc., Wilmington, Delaware.

of temperature is dependent upon the nature and concentration of the other dissolved ingredients. This is shown in Table 1.

Hydroxypropylcellulose is soluble in aqueous alcohol. The viscosity varies with the alcohol/water composition. The viscosity is usually greatest at a composition of 7 parts water to 3 parts alcohol by weight. This is illustrated for an ethanol water solvent system in Figure 3.

This type of viscosity curve is obtained whether the gum is added directly to the aqueous alcohol or whether it is first dissolved in either the water or the alcohol with the second solvent added afterwards.

The precipitation temperature from an alcohol and water solution is higher than that from an aqueous solution. The degree by which this temperature is increased depends upon the type and concentration of alcohol. Solutions containing 45% by volume of ethanol can be heated to boiling point without the hydroxypropylcellulose precipitating. Low concentrations of alcohol up to 25 to 30% have only slight effect on the precipitation temperature.

Propylene glycol performs in a manner similar to alcohol and precipitation temperature elevation follows a similar pattern.

3. Effect of pH

Aqueous solutions of hydroxypropylcellulose are usually stable over a pH range of 3 to 10. This permits its effective use in some food systems where low pH conditions prevail. Stability in low pH is greater than for other cellulose ethers, but over prolonged time periods, viscosity loss induced by hydrogen ion concentration can occur.

4. Viscosity Stability

While the physical stability of solutions is good, solutions can be degraded by acid-catalyzed hydrolysis, microorganisms, and by heat if dissolved oxygen is present. Solutions can be degraded by molds and bacteria; and therefore, if solutions are intended to be stored for a long period of time, a suitable preservative should be used. Resistance to enzyme degradation is generally better than that of cellulose gum due to the high level of substitution.

5. Surface Activity

Hydroxypropylcellulose is surface active. Aqueous solutions demonstrate reduced surface and interfacial tension. This property leads to its use as an emulsifying and whipping aid. These properties enable hydroxypropylcellulose to perform dual functions in oil and water emulsions and in aerated products.

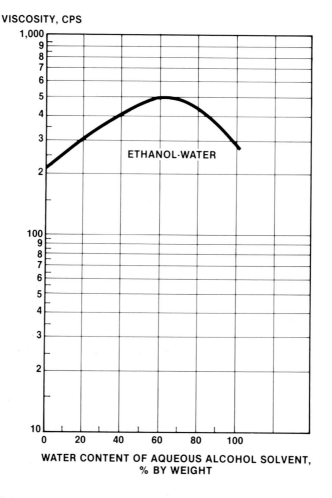

FIGURE 3.   Viscosity of aqueous alcohol solutions (Klucel G® at 2% concentration by weight).

C. Rheology of Aqueous Solutions

Solutions are generally smooth flowing and exhibit little structure or thixotropy. However, as solutions are non-Newtonian, under conditions of high rates of shear, they will show a temporary decrease in viscosity. The viscosity will revert to the original value when the shear force is removed. This effect is most apparent with high molecular weight grades.

D. Compatibility

In general terms, hydroxypropylcellulose has a good range of compatibility with both low molecular weight and polymeric organic materials. However, when used in aqueous systems, the solution is sensitive to high concentrations of other dissolved materials. The fact that hydroxypropylcellulose has hydrophilic as well as lipophilic properties tends to reduce its ability to hydrate in water. The presence of high concentrations of other dissolved materials can lead to a "salting out" of the dissolved hydrocolloid.

The compatibility of solutions with inorganic salts varies according to the salt. If relatively high concentrations of dissolved salts are present, the "salting out" effect can cause a finely divided precipitate to be formed. This effect generally results in decrease in viscosity and cloudiness in the solution.

Because of its hydroxypropyl substitutents, hydroxypropylcellulose tends to be more lipophilic in nature than other food grade cellulose derivatives. As a result, it is compatible with a wide range of anionic, nonionic, and cationic surfactants. The compatibility of solutions with surface active agents varies according to the particular type and concentration as well as with the grade and concentration of the hydroxypropylcellulose.

E. Film Formation

One of the most interesting properties of hydroxypropylcellulose is its ability to form films and sheets. It can be used in the formation of coatings on foods and protective film formation in some food applications. Klucel® films prepared by deposition from water solution have good barrier properties, tensile strength, and flexibility. They can act as very effective barriers to oil and fat.[1]

VII. FOOD APPLICATIONS

A. Whipped Toppings

Hydroxypropylcellulose demonstrates surface activity and also lipophilic properties. In addition, in common with other cellulose ether gums, it is an effective viscosifier and water binding agent. This combination of properties is used effectively in whipped toppings where the hydroxypropylcellulose stabilizes the emulsion, assists aeration and minimizes syneresis even after freeze-thaw cycling.

In ready to use whipped toppings, medium-viscosity hydroxypropylcellulose can be used in amounts varying from 0.2 to 0.5% of the final product, in combination with emulsifiers, a suitable fat, sugar, and flavors. The hydroxypropylcellulose and emulsifiers are normally dispersed in the aqueous phase using cold water to prevent precipitation of the hydroxypropylcellulose, followed by addition of sugar and melted fat. The resulting mixture is homogenized to produce a stable emulsion. The emulsion can be pasteurized, provided that homogenization of some kind is employed after heating to insure uniform distribution of hydroxypropylcellulose (as well as uniform oil droplet size). The emulsion should be cooled with agitation to facilitate redissolution of the hydroxypropylcellulose. Textural modifications can be achieved by adjusting the hydroxypropylcellulose concentration from the low end of the suggested range for a light, creamy texture to the high end (around 0.5%) to produce a more resilient, heavier texture. Fine tuning of texture can be achieved by adjusting overrun, fat type, emulsifier, and hydroxypropylcellulose levels.

Dry mix whipped toppings can also be produced using hydroxypropylcellulose, due to its cold solubility and emulsion-stabilizing properties. A simplistic model system can be achieved by using a medium-viscosity hydroxypropylcellulose together with commercially available spray-dried creaming agents and sugar. The basic system can be improved and modified by the use of milk solids and emulsifiers (Table 2). Stability of 2 to 3 days can be achieved by the stabilizing effect of the hydroxypropylcellulose.

B. Coatings and Film Forming

The film forming properties of hydroxypropylcellulose allow it to be used as a coating on a variety of food products. Klucel® (Hercules Inc.) can be deposited from an aqueous or polar organic solvent to form a clear flexible film which is impervious to fat or oil and which delays moisture and oxygen transfer.[1]

Although the technology of solvent removal can be difficult in some food operations, protective coatings can be applied to fried goods, cereals, nuts, and candy products to improve appearance and minimize fat or moisture pick up. In addition, hydroxypropylcellulose can be used as a glaze.

Klucel® coatings can be applied to foods by dipping, spraying, or in rotary coating pans. It should be remembered that Klucel® is a viscosifier, and therefore, to ensure maximum loading effect without developing too much viscosity in the coating solution, a low-viscosity

Table 2
INSTANT WHIPPED TOPPING
EXAMPLE

Klucel MF®	0.4%
Sucrose	27.0%
Dextrose	18.0%
Combined whipping agent/emulsifier[a]	54.0%
Colors and flavors	0.6%

Note: The final product is produced by mixing 50 g of dry mix with 120 g of cold milk using a domestic electric mixer, followed by whipping for 2 to 3 min until topping is light and fluffy.

[a]   Several commercially available types are suitable. In the above example Lamequick M, manufactured by Chemische Fabrik Grünau GmbH was used.

Courtesy of Hercules, Inc., Wilmington, Delaware.

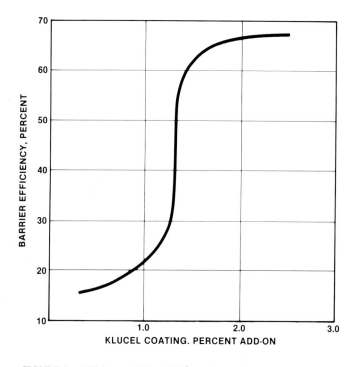

FIGURE 4.   Efficiency of Klucel LF® coatings to prevent moisture absorption by peanuts.

grade such as Klucel EF® would normally be used. A 10% solution of Klucel EF® in water or alcohol has a viscosity of 200 to 500 cps at room temperature. Additionally, as Klucel® is insoluble in water above 40 to 45°C (105 to 113°F), care should be used when drying aqueous coatings to avoid incomplete or irregular film formation that could occur if the temperature were too high.

Reference has already been made to the moisture and oxygen barrier properties of Klucel® films on foods. As an example, the moisture barrier efficiency for peanuts dipped in a 95% alcohol solution of Klucel® followed by air drying is shown in Figure 4.

In this example, moisture barrier efficiency is defined as:

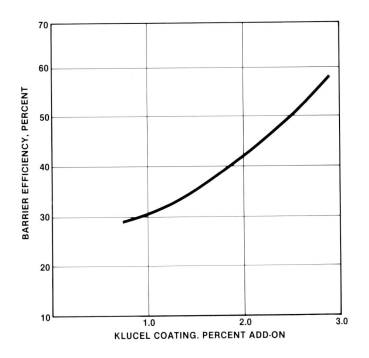

FIGURE 5. Efficiency of Klucel LF® coatings to prevent increase of peroxide value in peanuts.

$$\% \text{ barrier efficiency} = \frac{\%\text{moisture gain of uncoated nuts } - \%\text{moisture gain of coated nuts}}{\%\text{moisture gain of uncoated nuts}} \times 100$$

The coated and uncoated nuts were exposed to 80% relative humidity at 77°F for 24 hr.

Figure 5 shows graphically the oxygen barrier efficiency of Klucel® film determined from changes in the peroxide value of the oil expressed from the nuts after exposure to air at a temperature of 205°F in a forced air oven for 24 hr. The % barrier efficiency is expressed as follows:

$$\% \text{ barrier efficiency} = \frac{\text{PV of uncoated nuts } - \text{PV of coated nuts}}{\text{PV of uncoated nuts}} \times 100$$

A 2% by weight addition of Klucel® to the nuts indicates approximately 40% efficiency in preventing increase in peroxide value of the nuts and a 50 to 60% barrier efficiency against moisture pick up.

The property of preventing oil or fat migration is used in some candy products where peanut butter centers are coated with Klucel® prior to sugar coating to prevent oil seepage into the candy coating.

Klucel® can also be used with other coating materials such as starches, carboxymethyl-cellulose, and dextrins to improve the toughness and flexibility of films formed from these materials, which on their own, tend to be brittle.[1]

C. Salad Dressings

The use of Klucel® as a stabilizer in salad dressings has been suggested.[2] The properties of clarity of solution, cold solubility, and better acid and enzyme stability than cellulose gum would suggest Klucel® as a suitable stabilizer and viscosifier for clear oil-free dressings and marinades which contain vegetable pieces such as onion and garlic. Vegetables are often a source of cellulase enzymes. Cold processing to avoid expensive energy costs increases the risk of cellulase enzymes being present in the finished dressing or marinade.

D. Miscellaneous

The combined properties of polar organic solvent solubility, cold-water solubility, surface activity, thermoplasticity and film forming would suggest a variety of functional applications for hydroxypropylcellulose in the food and related industries.

The use of hydroxypropylcellulose to form edible water-soluble blow-molded containers has been reported.[3] A process for making thermoplastic shaped food products utilizing a mixture of hydroxypropylcellulose and a solid food product, processed under temperature and pressure sufficient to cause the mixture to soften and form shapes is described in U.S. Patent 3,769,029. Water is evaporated from the mixture to form a cohesive end product.[4]

In the alcoholic beverage industry in recent years, ready-to-drink creamy cocktails and cream whiskey products have become successful in the marketplace. One could anticipate that hydroxypropylcellulose might be a useful stabilizer in some of these types of products where alcohol and water solubility together with emulsion stabilizing properties may be needed.

REFERENCES

1. **Hercules Inc.**, *Klucel® for Edible Coatings*, Number VC 468, Wilmington, Del., 1981.
2. **Glicksman, M.**, *Gum Technology in the Food Industry*, Academic Press, New York, 1969, chap. 12.
3. **Anon.**, *Edible and Water Soluble Films*, Food Engineering International 1972, 2(8), 49—50.
4. **Ganz, A. J.**, U.S. Patent 3,769,029, 1973.

Chapter 4

METHYLCELLULOSE (MC) AND HYDROXYPROPYLMETHYLCELLULOSE (HPMC)

Joseph A. Grover

TABLE OF CONTENTS

I. BACKGROUND

Methylcellulose was one of the earliest synthetic substitutes for natural gums in food applications. In addition to use in the food industry, both methylcellulose and hydroxypropylmethylcellulose find broad application in a variety of industrial and pharmaceutical uses, including the construction, coatings, chemical, cosmetics, and textile industries. Early studies on etherification of cellulose were carried out by Suida[1] and, in particular, by Lilienfield,[23] who authored basic patents in the cellulose ether industry. Commercial manufacture of methylcellulose dates from the 1920s in Germany[4] and the 1930s in the U.S. Hydroxypropylmethylcellulose was produced in the U.S. starting in the late 1940s. Patents covering process of manufacture and composition of matter for hydroxypropylmethylcellulose were issued to Savage and assigned to the Dow Chemical Company.[5,6]

The cellulose ether gums are products of a chemical modification of cellulose,

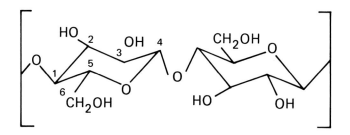

the most abundant organic chemical in the world. Through photosynthesis, billions of tons of cellulose are created each year and annual world consumption of cellulose for a myriad of applications is greater than 500 million tons.

As a result of the immense existing and potential supply of this starting material, the cellulose ethers, unlike the natural gums, are not subject to wide fluctuations in availability and price. The prices of the cellulose ethers have shown increases over the past decade corresponding to increases in costs of capital, raw materials, and energy. World capacity for methylcellulose and hydroxypropylmethylcellulose manufacture is estimated at greater than 100 million lb/year. This capacity may be readily increased with increases in long-term demand as there exist a number of world-wide manufacturers.

The use of methylcellulose and hydroxypropylmethylcellulose in foods is based upon the performance of these materials in imparting functional properties such as adhesion, emulsification, emulsion stabilization, viscosity control, gel formation, moisture retention, barrier properties, and freeze-thaw stabilization. The widespread acceptance of methylcellulose and hydroxypropylmethylcellulose in foods has been the subject of a number of publications and review articles. Of particular note for reference in the food areas are works by Glicksman,[7] Graham,[8] Krumel and Sarkar,[9] and Sarkar.[10] Data sheets which are very helpful are also available from manufacturers. Other publications which are highly informative regarding methylcellulose and hydroxypropylmethylcellulose for a wide variety of applications include those of Savage,[11] and Greminger and Savage.[12]

II. DESCRIPTION

Methylcellulose is described in the United States Pharmacopeia,[14] as "a methyl ether of cellulose containing not less than 27.5 percent and not more than 31.5 percent of methoxy (OCH$_3$) groups, calculated on the dried basis." Hydroxypropylmethylcellulose as approved for use as a food additive is described in the United States Pharmacopeia[14] under three monographs as hydroxypropylmethylcellulose 2910, 2906, and 2208. Further identification, information, and specifications are supplied in these references.

III. REGULATORY STATUS

Methylcellulose USP and hydroxypropylmethylcellulose USP,[14] meeting the specifications described in this publication, are approved as food additives under Title 21 of the Code of Federal Regulations.[15] Hydroxypropylmethylcellulose is now permitted in confectionary applications.[13] Methylcellulose USP, as described, is included in the list of food additives generally recognized as safe (GRAS), and is therefore not subject to the requirements of the Food Additives Amendment of 1958. Hydroxypropylmethylcellulose is permitted as a food additive in accordance with good manufacturing practice for use as an emulsifier, film former, protective colloid, stabilizer, suspending agent, or thickener.

Methylcellulose is labeled as such while hydroxypropylmethylcellulose may be so labeled or, alternatively, the term "carbohydrate gum" may be used.[16]

Among standardized foods, both methylcellulose and hydroxypropylmethylcellulose are permitted as optional emulsifying agents for French and salad dressing.[17,18] Hydroxypropylmethylcellulose is also permitted under the Standard of Identity for fruit sherbets and water ices.[19]

IV. MANUFACTURE

The general processes by which methylcellulose and hydroxypropylmethylcellulose are manufactured are described in the patent literature.[1,5,6] The various steps in manufacture may be broadly described as follows.

A. Alkali Cellulose Preparation

Cellulose, in the form of powder, chips, shreds, or sheets, is converted to alkali cellulose by treatment with sodium hydroxide. Molecular weight control of the product may be exercised at this point by controlled oxidation of the alkali cellulose. The generalized chemical reaction for alkali cellulose formation is shown in Equation 1, but the actual structure of the alkali cellulose has not been resolved.[20] All of the hydroxyl groups are available for reaction, although they vary in reactivity. The six and two positions are

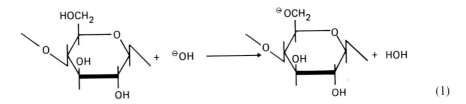

(1)

generally the most reactive, but the specific relative reactivity of the hydroxyl positions is a function of the reactants used and the reaction conditions.[21,24]

B. Etherification

Alkali cellulose is allowed to react with methyl chloride (and propylene oxide if hydroxypropylmethylcellulose is desired) to form the cellulose ethers. The etherifying reagents may react either simultaneously or in stages depending on loading sequences and reaction conditions.

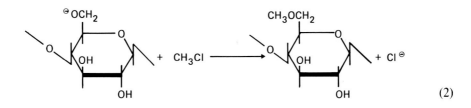

(2)

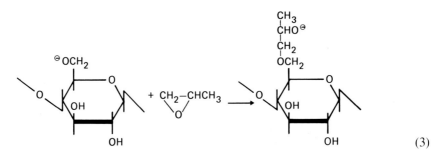

$$\tag{3}$$

Sodium hydroxide is consumed in the methylation reaction, but is required only as a catalyst in the hydroxypropylation reaction. Due to the presence of sodium hydroxide and water and the fact that an equilibrium between alkali cellulose and sodium hydroxide takes place, a substantial amount of by-product formation takes place (see Equations 4 to 8).

$$CH_3Cl + {}^{\ominus}OH \longrightarrow CH_3OH + Cl^{\ominus} \tag{4}$$

$$CH_3OH + {}^{\ominus}OH \rightleftarrows CH_3O^{\ominus} + HOH \tag{5}$$

$$CH_3Cl + CH_3O^{\ominus} \longrightarrow CH_3OCH_3 + Cl^{\ominus} \tag{6}$$

$$\underset{O}{\overset{CH_2-CHCH_3}{\diagup\diagdown}} + {}^{\ominus}OH \longrightarrow \underset{OH}{\overset{CH_2}{\mid}}-\underset{O^{\ominus}}{\overset{CHCH_3}{\mid}} \tag{7}$$

$$\underset{O}{\overset{CH_2-CHCH_3}{\diagup\diagdown}} + {}^{\ominus}OCH_3 \longrightarrow \underset{CH_3O}{\overset{CH_2}{\mid}}-\underset{O^{\ominus}}{\overset{CHCH_3}{\mid}} \tag{8}$$

Other reactions to form poly(propylene glycol) products and their methyl ethers also occur.

C. Purification

The unique thermogelation properties of methylcellulose and hydroxypropylmethylcellulose are used to simplify the purification of these products. Since these compounds are insoluble in hot water, they may be readily water-washed provided the temperature of the wash is maintained above the thermogelation temperature.[25] The by-products, sodium chloride, methanol, glycols, and glycol ethers, are all water-soluble and are readily removed to yield a highly purified product.

D. Drying and Grinding

The product resulting from the hot-water wash is in the form of a moist porous cake. This material is dried, normally using hot air as the drying medium. The dried product is ground to a powder and packaged.

E. Producers

In the U.S., the sole producer of methylcellulose and hydroxypropylmethylcellulose is the Dow Chemical Company, which started production under the trademark METHOCEL®*, in 1938. In the U.S., Dow Chemical produces METHOCEL® brand products at two locations,

---

* Trademark of The Dow Chemical Co.

one in Midland, Mich. and the other in Plaquemine, La. In Europe and Japan, there are a number of suppliers of methylcellulose and hydroxypropylmethylcellulose. These include Dow; Henkel & Cie, G.m.b.H. under the trade name CULMINAL; British Celanese under their registered name, CELACOL; Imperial Chemical Industries as METHOFAS; and Shin-Etsu Chemical Products, Ltd. under the trade name METALOSE. In addition, there are two companies which market methylcellulose, but not hydroxypropylmethylcellulose; Wolff and Co., under the trade name, WALSRODER; and Matsumoto Yishi as MARPOLOSE. Imperial Chemical Industries, Ltd., also markets a methylethylcellulose ether product under the name EDIFAS. A number of companies produce hydroxyethylmethylcellulose products. While hydroxyethylmethylcellulose products are not approved as direct food additives in the U.S., products with certain substitutions are allowed in a number of European countries. Methylcellulose and hydroxypropylmethylcellulose consumption levels world-wide are in the neighborhood of 70 million lb/year, and about 24 million lb/year in the U.S.[26] Although food and pharmaceutical uses account for only a fraction of this total, the use of these products in food applications is growing at a rate exceeding the growth rate of the cellulose ether business as a whole. In 1979, it was estimated that 2 million lb was consumed in food and pharmaceutical applications in the U.S. A growth rate in these applications corresponding to an anticipated use level of 3 million lb/year in 1984 was forecast.[26]

V. STRUCTURE

The cellulose repeating unit, the anhydroglucose moiety, has available three hydroxyl positions for etherification. In hydroxypropylmethylcellulose and methylcellulose, a variable number of these hydroxyl functional groups has been etherified.

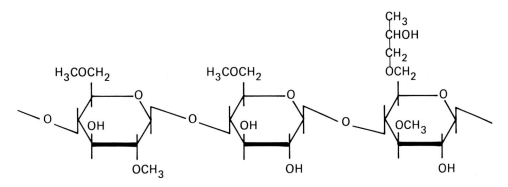

Hydroxypropylmethyl Cellulose

The terminology applied in describing these gums is to report the substitution as the degree of substitution, the DS, or as the molar substitution, the MS. The DS is defined as the average number of hydroxyl units per anhydroglucose unit which have been substituted. The MS is defined as the average number of molecules of substituent which have been substituted per anhydroglucose unit. In the case of hydroxypropoxyl substitution, it is possible to form poly(hydroxypropoxyl) chains since the added substituent generates a new hydroxyl group. Thus, it would be possible to have a cellulose ether with an MS of 2, i.e., average of 2 mol of hydroxypropoxyl functionally per mole of cellulose, but an average DS of 1 if the average chain length of the hydroxypropoxyl functionality is 2.

On the other hand, in the case of methoxyl substitution, the MS and DS must be identical, since no possibility exists for substitution of more than one methyl substituent per hydroxyl group.

The relative reactivity of the various hydroxyl groups on the cellulose molecule has been

Table 1
METHOCEL® PRODUCTS OFFERED BY
THE DOW CHEMICAL COMPANY IN
PREMIUM GRADES

Product	Viscosity mPa-S[a]	DS methoxyl	MS hydroxypropyl
A 15LV	15	1.6—1.9	0
A 4C	100	1.6—1.9	0
A 4M	4,000	1.6—1.9	0
E 5LV	5	1.8—2.0	0.20—0.31
E 15LV	15	1.8—2.0	0.20—0.31
E 50LV	50	1.8—2.0	0.20—0.31
E 4M	4,000	1.8—2.0	0.20—0.31
F 50LV	50	1.7—1.9	0.10—0.20
F 4M	4,000	1.7—1.9	0.10—0.20
K 3	3	1.1—1.6	0.10—0.20
K 35	35	1.1—1.6	0.10—0.30
K 100LV	100	1.1—1.6	0.10—0.30
K 4M	4,000	1.1—1.6	0.10—0.30
K 15M	15,000	1.1—1.6	0.10—0.30
K 100M	100,000	1.1—1.6	0.10—0.30

[a]   As a 2% aqueous solution.

the subject of a number of studies.[21-24] Since it is difficult in commercial processes to control the selectivity of the reagents for a given hydroxyl group (with one exception, see below), there has been little effort to attempt to correlate functional behavior with the distribution of substituents on the individual anhydroglucose units.

The one exception, referred to above, is that by choosing between sequencing the addition of methyl chloride and propylene oxide or adding these reagents simultaneously in the manufacturing process, a certain degree of control of location of the substituent on the anhydroglucose unit is achieved.[27] Manipulation of substitution in this manner is not believed to be of great significance with products which currently find use in food applications. The effects of changing the absolute level of substitution of methoxyl and hydroxypropoxyl functionalities, however, has a profound effect on a number of physical and chemical properties of these gums. Water retention properties, sensitivity to electrolytes and other solutes, dissolution temperatures, gelation properties, and solubility in nonaqueous systems are all significantly affected by variations in methoxyl and hydroxypropoxyl substitution within the range permitted for food and drug applications. The range of methoxyl substitution permitted under the CFR for methylcellulose is 27.5 to 31.5%. This corresponds to a DS range of 1.64 to 1.92. For hydroxypropylmethylcellulose the range is 3.0 to 12.0% of hydroxypropoxyl functionality with 19.0 to 30.0% of methoxyl substitution. These ranges correspond to an MS range of 0.073 to 0.336 for hydroxypropoxyl and a DS range of 1.11 to 2.03 for methoxyl. Commercially available products as produced by the various suppliers fall within these ranges. An example of the broad range of products offered is shown in Table 1.

An important feature for commercial applications of the cellulose ethers is that they are available in a wide range of molecular weights. Producers of methylcellulose and hydroxypropylmethylcellulose have recently developed technology which has allowed them to extend the range of product molecular weights available both at higher and lower molecular weight ranges. The relationship of 2% aqueous solution viscosities and the average molecular weight has been determined and reported. This relationship is shown in Figure 1 and described in Table 2.

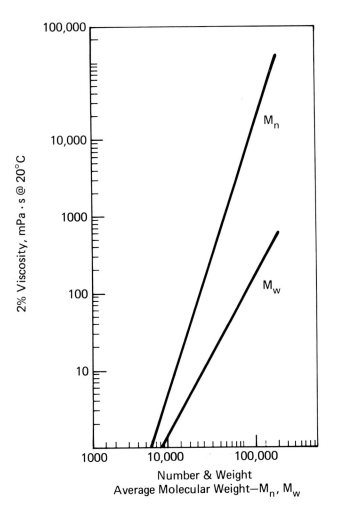

FIGURE 1. Molecular weight/viscosity correlation. (Reprinted by permission of the Dow Chemical Company.[28])

In theory it is possible to determine molecular weight of water-soluble polymers by gel permeation chromatography as has been done for organosoluble cellulose derivatives.[29,30] However, the very large hydrated size of the water-soluble cellulose ethers coupled with extreme sensitivity of the hydrated polymer volume to the presence of small quantities of electrolytes or other interacting species has not allowed broad development of this technique. The use of gel permeation chromatography to determine molecular weight of polymers in aqueous solution is currently a topic of research interest.

VI. PROPERTIES

A. Solubility

Cellulose itself is highly insoluble in most solvents due to the very high level of intramolecular hydrogen bonding in the cellulose polymer. A high degree of crystallinity is also present which reduces solubility as well. In order for dissolution of the polymer to take place, the crystallinity and intramolecular hydrogen bonding must be reduced.

In the case of the cellulose ethers, this is accomplished by etherification. In effect, the ability of the cellulose molecule to hydrogen bond is reduced, but the capacity of the polymer

Table 2
VISCOSITIES OF METHYCELLULOSE OF VARIOUS
MOLECULAR WEIGHTS

Viscosity grade 2%, 20°C, mPa·S	Intrinsic viscosity (η), dℓ/g at 20°C	Number average DP_n	Number average molecular weight, M_n
5	1.2	53	10,000
10	1.4	70	13,000
40	2.05	110	20,000
100	2.65	140	26,000
400	3.90	220	41,000
1,500	5.7	340	63,000
4,000	7.5	460	86,000
8,000	9.3	580	110,000
15,000	11.0	650	120,000
19,000	12.0	750	140,000
40,000	15.0	950	180,000
75,000	18.4	1,160	220,000

From *Encyclopedia of Polymer Science and Technology,* 3, Interscience, New York, 1965, 504. With permission.

molecule to be hydrated is not eliminated. In addition, replacement of hydroxyls by bulkier ether groups sterically inhibits close fitting and crystallization of cellulose molecules. As the level of etherification is increased, the cellulose molecule proceeds from a species which is swollen by alkali to a partial ether which is soluble in dilute alkaline solution. This takes place at a methoxyl substitution of about 16 to 22.5%, corresponding to a DS of about 0.9 to 1.3. Further etherification results in water solubility being obtained at a DS of about 1.4, equivalent to a methoxyl content of about 24%. Swelling and partial solubilization in various organic solvents is noted at a DS of 2.1, corresponding to 34% methoxyl content. Water solubility with good organic solubility is found at about 36% OCH_3, or a DS of 2.25. Water solubility is eventually lost altogether, and the methylcellulose becomes soluble in organic solvents at a DS of about 2.6, in the range of 40% or higher methoxyl substitution. The effect of hydroxypropoxyl substitution is to broaden the range of both water and organic solvent solubility.

The most organosoluble of the methylhydroxypropylcellulose ethers which are permitted for food applications are those which have nearly the maximum permitted substitution of both substituents. Such a product is typefied by METHOCEL® E which may have a DS of as high as 2.03 methoxyl and an MS of 0.336 hydroxypropoxyl. This material is soluble in a number of solvents at elevated temperatures and in aqueous/organic mixed solvents at room temperature as may be seen in the data presented in Table 3.

When preparing aqueous solutions of the cellulose ethers, care must be taken to prevent formation of lumps. When properly dispersed, clear solutions may be obtained in a matter of minutes. If, due to improper solution makeup techniques, lumps are allowed to form, it often requires many hours to effect complete dissolution.

There are three commonly used techniques to promote full dispersion and uniform wetting of methylcellulose particles to promote rapid and complete dissolution. The first of these takes advantage of the unique thermogelation properties of this family of gums. In this method, the cellulosic gum is dispersed in water which has been heated to a temperature above the gelation temperature of the gum; i.e., the temperature above which the compound is insoluble in water. Normally, the dispersion is accomplished using about one quarter to one half of the total water required. Once the polymer is fully wetted and dispersed in the

Table 3
REPRESENTATIVE SOLVENTS FOR METHOCEL® E
BRAND CELLULOSE ETHER AT ELEVATED
TEMPERATURES

Compound	Boiling point, °C	Solubility point, °C	Degree of solubility[a]
Glycols			
Ethylene glycol	197.3	158	C
Diethylene glycol	244.8	135	C
Propylene glycol	188.2	140	C
1,3-Propanediol	214	120	C
Glycerine	290	260	P
Dowanol® EE	134.7	120	C
Ethylene glycol			
Ethyl ether			
Dowanol® TPM	242.4	160	P
Tripropylene glycol			
Methyl ether			
Esters			
Ethyl glycolate	160	110	C
Glyceryl monoacetate (acetin)	127/3 mm	100	C
Glyceryl diacetate (diacetin)	123—133/4 mm	100	C
Amines			
Monoethanolamine	170—172	120	C
Diethanolamine	268—269	180	C

[a]   C: Completely soluble; P: partially soluble

Reprinted by Permission of The Dow Chemical Company, 1978.

hot water, the rest of the water is added as ice or cold water while agitation is continued. This addition cools the dispersion and full dissolution occurs rapidly (Figure 2).

A second technique which is often used when a number of other dry ingredients are called for in the formulation is to dry blend the gum with the rest of the dry ingredients. When liquid is eventually added, lump formation is prevented and hydration takes place normally.

A third method involves the wetting of the cellulosic material with nonaqueous liquid ingredients, e.g., propylene glycol, followed by the addition of water which then promotes lump-free dissolution with adequate agitation.

The choice of technique involves consideration not only of the convenience of incorporation, but also the function of the cellulosic gum in the system to which it is being applied. Experimentation may show that the procedure for achieving dispersion may affect the performance of the gum in the system due to the degree of hydration which is achieved. This is particularly true in food formulations due to the limited amount of water which is present and the competition of the various hydrophilic ingredients for the available water.

B. Viscosity

The eighth root of measured methylcellulose and hydroxypropylmethylcellulose solution viscosities shows a linear relationship with concentration which is illustrated in Figures 3 and 4. This is an empirically derived relationship. Since solutions of these cellulose ethers, other than very dilute solutions, are highly pseudoplastic, the observed viscosity is highly dependent on the viscometer used. This property will be discussed in more detail later. For very dilute solutions of certain cellulose ethers, the Martin equation has been shown to be valid with a Martin k of 0.191 for hydroxyethylcellulose.[31]

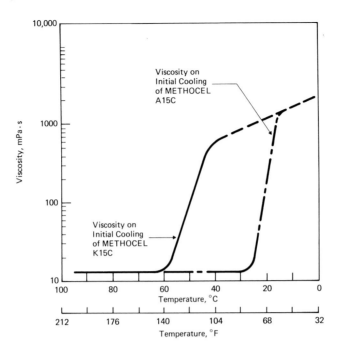

FIGURE 2.   Viscosity development of METHOCEL® K15C and METH-OCEL® A15C brand products slurried at 2% in hot water. (Reprinted by permission of the Dow Chemical Company.[28])

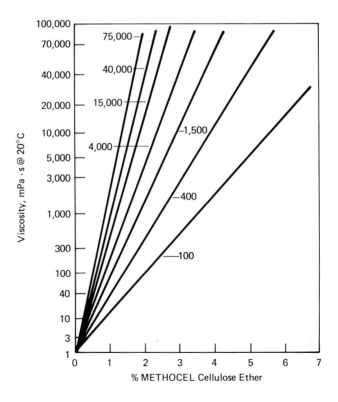

FIGURE 3.   Viscosity-concentration chart for high viscosity METHO-CEL® products. (Reprinted by permission of the Dow Chemical Company.[28])

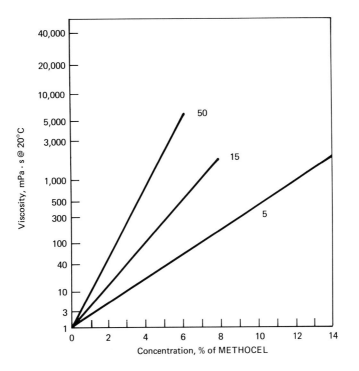

FIGURE 4.   Viscosity-concentration chart for low viscosity METHO-CEL® products. (Reprinted by permission of the Dow Chemical Company.[28])

$$\log(\eta_{sp}/c) = \log[\eta] + k[\eta]c,$$

It is expected that this equation will also hold for methylcellulose and hydroxypropylmethylcellulose, although a different k value may be found.

The viscosity of an aqueous solution of methylcellulose or hydroxypropylmethylcellulose will initially decrease upon heating. However, when the gel temperature for the particular gum is reached, the solution will gel, causing a very substantial increase in measured viscosity and the development of a yield value. This gelation phenomenon, as shown in Figure 5, will be discussed in detail separately.

Since these gums are nonionic, their solutions are relatively insensitive to changes in pH. In this respect, they differ significantly from both cationic and anionic water-soluble polymers which normally show dramatic changes in solubility and solution viscosity near their isoelectric points. In common with many polysaccharides, the cellulose ethers undergo hydrolysis at extremes of pH (less than pH 3 or greater than pH 11), particularly in strong acid solutions. In vinegar-containing systems, they are reasonably stable, but at lower pH values, hydrolysis possibilities must be considered. The hydrolysis rates of some methylcelluloses in hydrochloric acid solution are shown in Figure 6. It is seen that the effect of level of substitution on hydrolysis rate is only minor.

The nonionic character of methylcellulose and hydroxypropylmethylcellulose which is responsible for the stability of viscosity to changes in pH also result in solutions which are highly tolerant to electrolytes. Although added electrolytes do cause gradual viscosity loss with increasing concentration, these gums are less sensitive to electrolytes than are ionic water-soluble polymers. However, the presence of certain surfactants, in particular the sulfates and sulfonates, has profound effects on the viscosity of cellulose ether solutions. The effects of some additives and of some of these surfactants are tabulated in Tables 4 and 5.

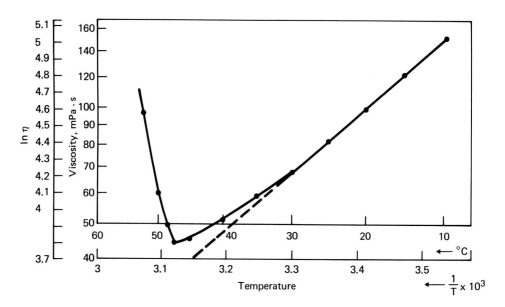

FIGURE 5. Viscosity-temperature relationship for 2% aqueous solution of METHOCEL® A100 methylcellulose at a shear rate of 86 sec[-1]. (Reprinted by permission of the Dow Chemical Company.[28])

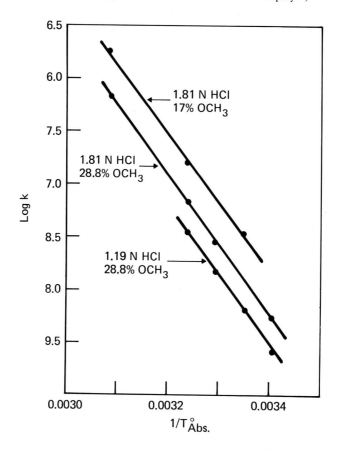

FIGURE 6. Arrhenius plots for hydrolysis of methylcelluloses in HCl. (Reprinted by permission of the Textile Institute.[32])

Table 4
GRAMS OF ADDITIVE TOLERATED BY 100 cc 2%
SOLUTION WITHOUT SALTING OUT

| | METHOCEL® | | | | | | |
Additive	A15	A4M	F50	F4M	K100	K4M	J5M
NaCl	11	7	17	11	19	12	10
$MgCl_2$	11	8	35	25	40	39	22
Na_2SO_4	6	4	6	4	6	4	3
$Al_2(SO_4)_3$	3.1	2.5	4.1	3.6	4.1	3.6	2.7
Na_2CO_3	4	3	5	4	4	4	3
Na_3PO_4	2.9	2.6	3.9	3.5	4.7	4.3	2.5
Sucrose	100	65	120	80	160	115	100

Reprinted with permission of the Dow Chemical Company, 1974.

In general, the effects of additives on viscosity must be determined experimentally for particular systems, since it is not possible to predict with accuracy the magnitude or direction of these effects.

C. Solution Rheology

As previously stated, the rheological behavior of methylcellulose and hydroxypropyl-methylcellulose solutions is dependent upon the concentration of the dissolved polymer hydrocolloid. Other factors which affect the solution rheology include the molecular weight of the polymer, the solution temperature, and as has been noted, the presence of other solutes. Solutions of methylcellulose and hydroxypropylmethylcellulose generally exhibit pseudoplastic behavior, i.e., the solution viscosity decreases with increasing rate of shear. However, at very low shear rates, solutions of these gums exhibit Newtonian properties. The degree of pseudoplasticity increases with increasing concentration and increasing molecular weight. Therefore, the shear rate above which pseudoplastic behavior becomes apparent decreases as higher molecular weight polymers are used and as the solution concentration is increased, as is illustrated in Figures 7 and 8.

In addition to the above shear rate dependence, certain systems containing methylcellulose and hydroxypropylmethylcellulose demonstrate a thixotropic behavior, i.e., the viscosity will decrease with time at constant shear. This property is often exhibited by methylcellulose solutions containing a substantial amount of undissolved solids.

D. Thermogelation
1. Definition

The unique property of the formation of completely reversible gels of methylcellulose and hydroxypropylmethylcellulose upon heating their solutions is responsible for many of the food use applications of these products.[10,34] The applications include, among others, binders for reconstituted meat and vegetable products, foam stabilization, film formation in fried foods, stabilization of fruit pie fillings during baking, and dough strengthening during the baking of low-gluten bakery products.[16,35]

The phenomenon of thermogelation of methylcellulose products is graphically demonstrated in Figure 9 showing the effect of heating on the viscosity of a methylcellulose solution. The incipient gelation temperature (IGT) is defined as the temperature at which the viscosity reaches a minimum. Once gelation begins, viscosity build takes place due to intermolecular and intramolecular association. Since the dehydration and association leading to gelation are time-dependent processes, the rate of temperature change will have an effect upon the gelation

Table 5
EFFECT OF ADDITIVES ON VISCOSITY OF 1% SOLUTIONS OF METHOCEL® CELLULOSE ETHER

Trademark	% Additive	Producer	Description	Increase in viscosity of METHOCEL® brands			
				A	E	F	K
Conco AA5-35S	1	Continental Chemical Co.	Sodium dodecyl benzene sulfonate	227	1863	908	36
Conco sulfate EP	1	Continental Chemical Co.	Diethanolamine lauryl sulfate	145	104	94	53
Miranol® C2M conc.	1 / 25	Miranol Chemical Co.	Dicarboxylated imidazoline derivative of coconut fatty acid	0 / −80	15 / 75	1 / —	−1 / −2
Miranol® L2MSF	1 / 10	Miranol Chemical Co.	Dicarboxylated imidazoline derivative of tall oil fatty acid	11 / −58	6 / −81	11 / —	8 / −68
Miranol® 2MCT modified	1 / 10	Miranol Chemical Co.	Polyoxyethylene (3)tridecyl sulfate salt of a dicarboxylated imidazoline derivative of coconut fatty acid	10 / 41	0 / 34	5 / —	4 / 40
Miranol® HM conc.	1 / 10	Miranol Chemical Co.	Monocarboxylated imidazoline derivative of lauric acid	8 / 30	−1 / 30	5 / —	4 / 35
Polystep® B-11	1	Stepan Chemical Co.	Ammonium lauryl ethoxylate (4) sulfate	36	50	57	42
Quaternary 0	1	Geogy Chemical Corp.	Quaternary ammonium imidazoline derivative	26	22	22	18
Span® 60	1	ICI Americas	Sorbitan monostearate	104	100	124	105
Teepol® 610	1	Shell Chemicals UK Ltd.	Secondary sodium alkyl sulfonate	184	36	122	132
Triton® CQ 400	1	Rohm & Haas Co.	Stearyl dimethyl benzyl ammonium chloride	109	71	149	168
Tween® 20	1	ICI United States	Polyoxyethylene (20) sorbitan monolaurate	5	−8	−10	2
Ultrawet® 30 DS	1	ARCO Chemical Co.	Sodium linear alkylate sulfonate	2080	5845	2798	107

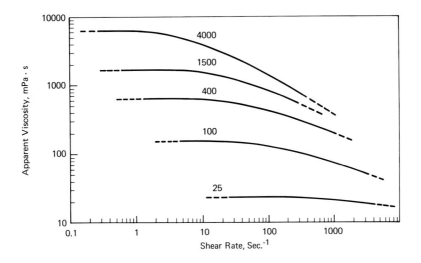

FIGURE 7.   Apparent viscosity vs. shear rate curves for 2% aqueous solutions of METH-OCEL® at 20°C. (Reprinted with permission of the Dow Chemical Company.[28])

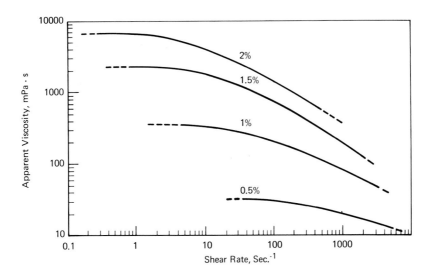

FIGURE 8.   Apparent viscosity vs. shear rate curves for aqueous solutions of a 4000 mPa S grade METHOCEL® product at different concentrations at 20°C. (Reprinted with permission of the Dow Chemical Company.[28])

temperature. The effect of the equilibration time between associated and hydrated forms leads to an hysteresis effect upon cooling the gel as is shown in Figure 9.

Some other terms which are important to a discussion of thermogelation are defined below and are illustrated in Figure 10.

The incipient precipitation temperature (IPT) is defined in reference to the light transmission of a methylcellulose solution. These gels are translucent rather than transparent and the loss of light transmission may be taken as a measure of thermogelation of the polymer in solution. If the solution is heated at 0.25°C/min, the temperature at which the light transmission at 545 nm has been reduced to 97.5% of the initial value, is defined as the IPT. The cloud point (CP) is the temperature at which the light transmission is reduced by 50% while heating at 0.25°C/min. Neither the IPT, which is an indication of the temperature at which the high molecular weight and/or the higher substituted fractions begin to precipitate

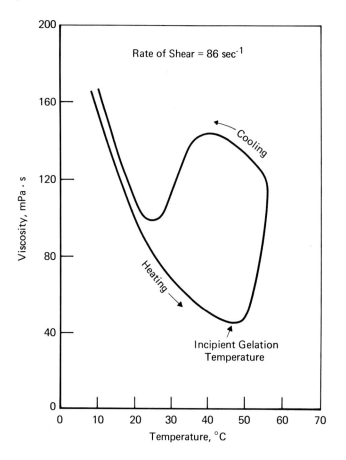

FIGURE 9.   Gelation of 2% aqueous solution of METHOCEL® A100 on heating at 0.25°C/min. (Reprinted with permission of the Dow Chemical Company.[28])

from solution, nor the CP accurately reflect the average molecular weight of the sample. The average molecular weight is better estimated by reference to the solution viscosity and the rate of increase in viscosity with increases in gum concentration.

2. Effects of Gum Concentration

A comparison of the effects of methylcellulose concentration upon the phenomena of IPT, CP, and IGT is very interesting in the contrasts displayed as shown in Figure 11. Both the IPT and the CP show a rapid decrease with increasing concentration at concentrations less than 1%, but tend to level out as the concentration is increased to the 3 or 4% level. The IGT, on the other hand, shows a linear decrease over the range of 2 to 6%. In the case of METHOCEL® F50 hydroxypropylmethylcellulose, as shown in this figure, it is indicated that at concentrations below about 6.5%, the solution becomes turbid before gelation takes place, while above this concentration a clear gel should form. It has been reported that with certain methylcellulose solutions, it is possible to obtain clear gels at room temperature. Normally, however, such high concentrations are required to achieve this behavior that solution makeup is difficult and viscosity levels obtained are so high that this behavior is not easily observed.

3. Factors Affecting Gel Strength

As was described earlier, the gelation phenomenon is time dependent. The rate of gel

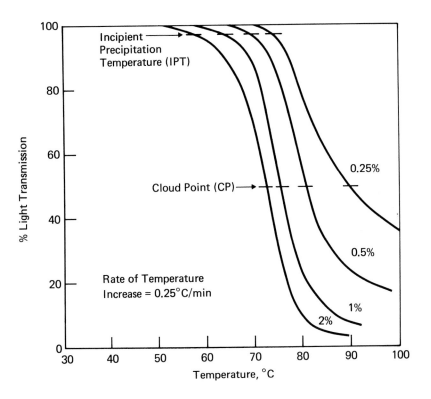

FIGURE 10.   An illustration of the change of 545 nm light transmission for aqueous solutions of hydroxypropylmethylcellulose sample, having a DS = 1.45 and MS = 0.11, as a function of temperature at different concentrations. (Reprinted by permission of the Dow Chemical Company.[10])

strength development for METHOCEL® A4C is shown in Figure 12 for a given set of experimental conditions.

Higher temperatures or the presence of electrolytes promote more rapid gelation due to the more rapid dehydration of the gum. In addition, the gel strength is increased by the presence of electrolytes. Syneresis, which is due to shrinkage of the gel and extreme dehydration, takes place more rapidly at higher temperatures and in the presence of electrolytes and also results in an increase in gel strength. The dependence of the IPT, CP, and gel strength of a methylcellulose sample upon sodium chloride concentration is shown in Figure 13.

Other factors which have been shown to substantially affect gelation properties of methylcellulose gums include the concentration of the gum, the molecular weight, the type of substituent, and the degree of substitution. In Figure 14, the dependence of gel strength upon the molecular weight is shown for methylcellulose gels. The gel strength increases with increasing molecular weight up to an average molecular weight of approximately 40,000 which corresponds to a nominal 2% solution viscosity of 400 mPa-S. Above this molecular weight, no further substantial increases in gel strength accompany increases in molecular weight.

An alternative way to form even stronger, more rigid gels is simply by increasing the concentration of the gum, as may be seen from Figure 15, or to have present an electrolyte or other solute as was shown in Figure 13. The increase in gel strength with increasing gum concentration is dramatic and may be useful when stronger, reversible gels are desired.

The effect upon gelation properties of combining hydroxypropoxyl substitution with methoxy substitution is also discussed by Sarkar in his comprehensive paper. He has shown that

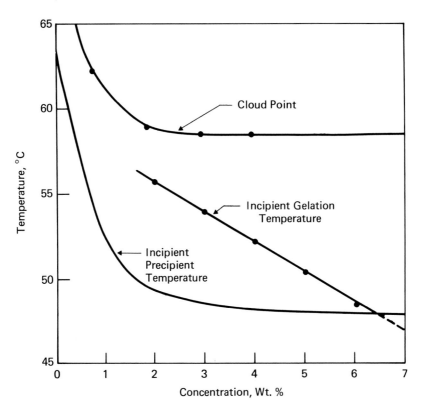

FIGURE 11.   Temperature of sol-gel transformation for aqueous solutions of METHOCEL® F50 as a function of concentration. (Reprinted by permission of the Dow Chemical Company.[10])

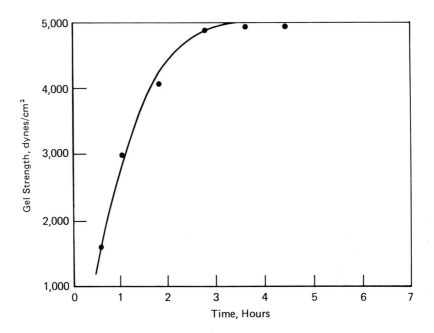

FIGURE 12.   Rate of gel strength development for a 2% solution of METHOCEL® A4C upon heating at 65°C. (Reprinted with permission of the Dow Chemical Company.[10])

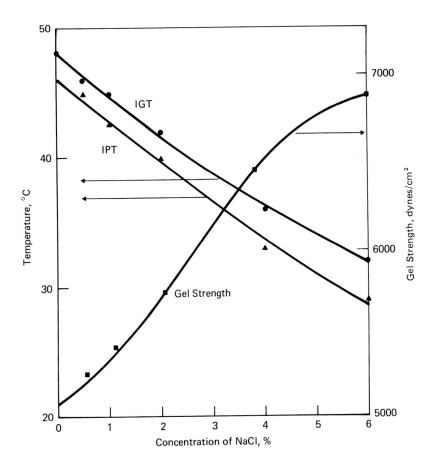

FIGURE 13. Incipient gelation temperature (IGT), incipient precipitation temperature (IPT), and gel strength of a 2% solution of METHOCEL® A15C methylcellulose as a function of NaCl concentration. (Reprinted by permission of the Dow Chemical Company.[10])

the IGT is raised by the addition of low levels of hydroxypropoxyl substitution (see Figure 16) and that the sensitivity of the IGT to concentration is reduced by increasing the amount of hydroxypropoxyl substitution at a relatively constant level of methoxyl substitution as may be seen from Figure 17. Accompanying the raising of the IGT with added hydroxypropoxyl substitution, there is a large reduction in the strength of the gel formed. In Figure 18, it is seen that a drop in gel strength to one fifth of the original level is obtained as the hydroxypropoxyl content is increased from 0 to an MS of approximately 0.17 at an essentially constant methoxyl DS of about 1.75. The dependence of gel strength phenomena upon this collection of factors allows substantial freedom in adjusting gel strength and texture while maintaining other properties, such as binding strength, film formation, or solution rheology, at a desired level.

The effects of various ionic and nonionic additives on gelation temperature are shown in Table 6 and in Figures 19 and 20. It is important to food applications that it is possible, by suitable selection, to either increase or decrease the gelation temperature. The normal effect of these additives upon gel strength is that those which raise the gelation temperature tend to soften the gel while those which lower the gel temperature make the gel firmer.

E. Compatibility With Other Hydrocolloids

The effects of electrolytes, pH changes, and certain surfactants have already been dealt with earlier in this chapter in discussing effects of additives on viscosity. In addition to the

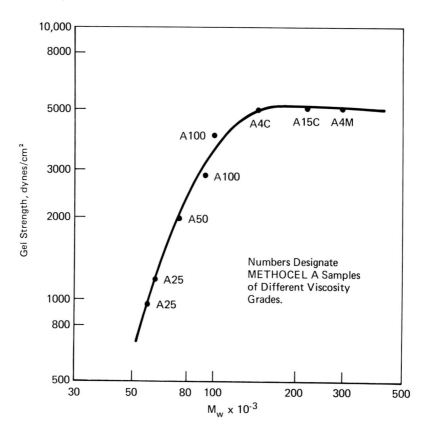

FIGURE 14.   Gel strength of 2% aqueous methylcellulose gels after 4 hr at 65°C as a function of molecular weight. (Reprinted by permission of the Dow Chemical Company.[10])

tolerance of solutions of methylcellulose gums to such compounds, they are also compatible with a substantial number of synthetic and natural water-soluble polymers. These include acacia, gum arabic, tragacanth, starch, starch ethers, carragheenan, the alginates, poly(vinyl alcohol), and xanthan gum as well as many others (see Table 7).

The limits of compatibility are affected by the structure of the added hydrocolloid and the molecular weights of the gums. Normally, lower molecular weight methylcellulose gums are more compatible than are higher molecular weight examples. In addition to simple compatibility, synergistic viscosity increases are sometimes exhibited by mixed gum systems such as methylcellulose with starch ethers or carboxymethylcellulose.[25]

In general, the behavior of mixed gum systems in food applications must be determined experimentally, since the effects of other components in the total product system may greatly alter the effects seen in a pure water system.

In cases in which fillers or solids are combined with methylcellulose solutions, no salting out or precipitation is observed. In fact, low molecular weight methylcellulose and hydroxypropylmethylcellulose may be used as dispersants in such systems. Again, the use of gum combinations including these products often leads to improved stabilization of emulsions and solids suspensions when compared to that obtained for either gum alone. Performance of gum combinations must normally be determined experimentally although the manufacturers of the cellulose ethers have published some recommended gum systems in their product literature.

The effects of additives on solutions of methylcellulose and hydroxypropylmethylcellulose may best be predicted by bearing in mind the expected effects of the additives on the

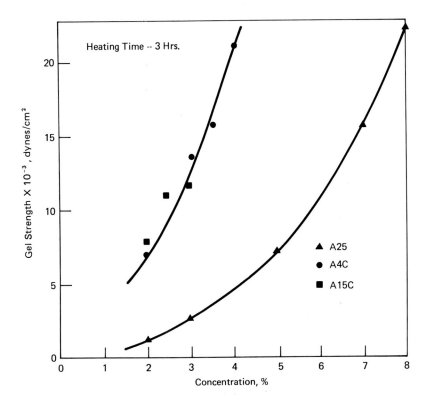

FIGURE 15. Gel strength of METHOCEL® A methylcellulose of different viscosity grades at 65°C as a function of concentration. (Reprinted with permission of the Dow Chemical Company.[36])

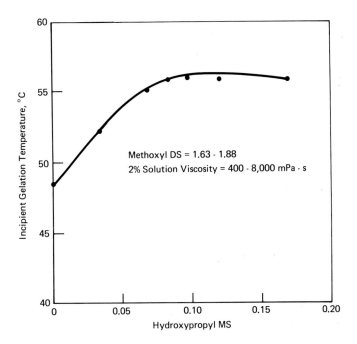

FIGURE 16. Incipient gelation temperatures of hydroxypropylmethyl-cellulose as a function of hydroxypropyl molar substitution. (Reprinted by permission of the Dow Chemical Company.[10])

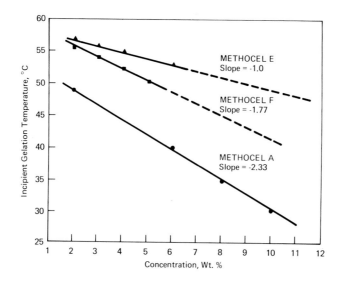

FIGURE 17. Incipient gelation temperature of different METHOCEL® products as a function of concentration. (Reprinted with permission of the Dow Chemical Company.)

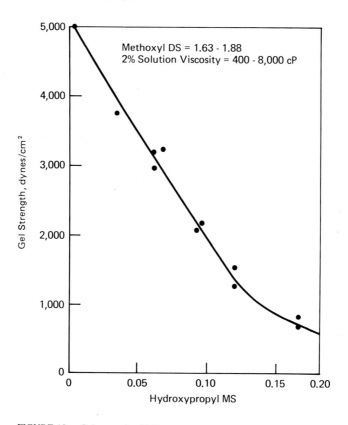

FIGURE 18. Gel strength of 2% aqueous hydroxypropylmethylcellulose gels after 4 hr at 65°C as function of hydroxypropyl molar substitution. (Reprinted with permission of the Dow Chemical Company.[10])

Table 6
**EFFECT ON GELATION TEMPERATURE NOTED WITH ADDITIVES TO 2%
SOLUTIONS OF METHOCEL® CELLULOSE ETHER**

Additive		METHOCEL® A15C		METHOCEL® F15C[a]		METHOCEL® K4M	
Compound	%	°C	°F	°C	°F	°C	°F
Control (no additive)		50	122	63	145	85	185
NaCl	5	33	91	41	105	59	138
MgCl$_2$[b]	5	42	107	52	125	67	153
FeCl$_3$	3	42	107	53	127	76	169
Na$_2$SO$_4$	5	Salted out		Salted out		Salted out	
Al$_2$(SO$_4$)$_3$	2.5	Salted out		45	113	48	118
Na$_2$CO$_3$[b]	5	Salted out		Salted out		Salted out	
Na$_3$PO$_4$	2.0	32	89	42	107	52	125
Sucrose	5	51	124	66	151	84	183
Sucrose	20	44	111	59	138	61	142
Sorbitol	20	30	86	46	115	48	118
Glycerine[b]	20	34	93	60	140	65—70	149—158
Ethanol	20	>75	>167	>75	>167	>75	>167
Polyethylene glycol 400[b]	20	52	126	>80	>176	>80	>176
Propylene glycol[b]	20	59	138	>80	>176	>80	>176

Note: Of the compounds in the table, sucrose, ethanol, and the two polyglycols raise the gelation temperature. Unlisted additional compounds that raise the thermal gel temperature include Armac®[c], Armac®HDT[c], Hyamine 1622,[d] alkali metal thiocyanates, and urea.

[a] A special viscosity grade made by blending.
[b] The Dow Chemical Company
[c] Armour and Company
[d] Rohm & Haas Company

Reprinted by permission of the Dow Chemical Company, 1978.

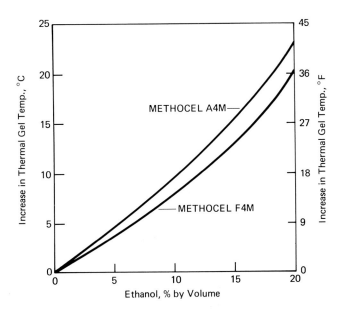

FIGURE 19.   Effect of ethanol on thermal gel temperature (2% solutions).
(Reprinted by permission of the Dow Chemical Company.[28])

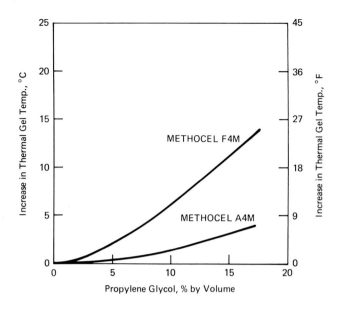

FIGURE 20.   Effect of propylene glycol on thermal gel temperature (2% solutions). (Reprinted with permission of the Dow Chemical Company.[28])

distribution of water in the system. Addition of components which will reduce water availability due to hydration or adsorption will act to increase the effective viscosity of the gum selected and will also tend to decrease the gel point. Compounds capable of hydrogen bonding to the cellulose ether may increase the gel point. Sodium thiocyanate, potassium thiocyanate, and urea act to increase the gel point rather than to decrease it.

F. Film Formation

Methylcellulose and hydroxypropylmethylcellulose are capable of forming films when cast from solution. Films may also be formed by rolling or extruding the gum after plasticizing with agents such as propylene glycol. Some properties of unplasticized films of methylcellulose and hydroxypropylmethylcellulose cast from water are shown in Table 8. Film physical properties such as tensile strength, elongation, and permeability are highly sensitive to moisture content of the film.

It is possible to use cellulose ether films as water-soluble packaging materials,[7] and films plasticized with plasticizers such as propylene glycol may even be heat sealed at temperatures as low as 130°C.

Other uses of the film-forming properties of methylcellulose films utilize the resistance of these films to oils and fats. Coating of foods with a methylcellulose containing batter or dip prevents oil absorption and helps prevent loss of moisture during deep frying. The inclusion of methylcellulose in extruded food products both improves glossiness and helps prevent crumbling. It is also possible to prevent loss of volatile components by encapsulating with a methylcellulose film as, for example, in the encapsulation of flavor oils.[38]

VII. FOOD APPLICATIONS

Methylcellulose and hydroxypropylmethylcellulose continue to find new applications and broader use in the food industry. These gums, due to the many functional properties which they embody, are capable of filling needs as alternative products, natural gum replacements, or more cost-efficient gum replacements in some cases. The most important food applications on a volume basis and the functional properties which are most significant to these appli-

Table 7
INTERACTION OF HPMC WITH OTHER HYDROCOLLOIDS IN AQUEOUS SOLUTION

Low viscosity HPMC	High viscosity HPMC	Guar	Low viscosity CMC	Gelatin	mPa-S at 10 sec⁻¹	Haze point (°C)	Gelation temperature (°C) On cooling	On heating	Comments
	0.5	—			24		None	—	
	1.0	—			190			—	
	2.0	—			3,800			56	
	—	0.2			74			—	
		0.6			1,210			—	
	0.5	0.2			110			—	
	0.5	0.6			980			58	
	1.0	0.7			750			58	
2.0			—		39.2		None	55	Structured gel
4.0			—		330			56	Structured gel
—			2.0		24.2			—	No gel
—			4.0		190			—	No gel
—			6.0		1,838			—	No gel
2.0			1.0		—	50		58	Poor gel structure
2.0			2.0		—	47		53	Poor gel structure
2.0			4.0		—	42		47	Poor gel structure
2.0			6.0		—	—		42	Poor gel structure
	2.0		—		4,700		None		Used high-viscosity CMC 7H4F
	—		2.0		34,300				
	1.0		1.0		23,250				
—				2.0			25	None	—
—				4.0			26—27	None	—
2.0				—			None	55	
4.0				—			None	56	
0.5				2.0			27	65	Hazy solution
1.0				2.0			28	68	Compatible

Table 7 (continued)
INTERACTION OF HPMC WITH OTHER HYDROCOLLOIDS IN AQUEOUS SOLUTION

Gum concentration					Haze point (°C)	Gelation temperature (°C)		Comments
Low viscosity HPMC	High viscosity HPMC	Guar	Low viscosity CMC	Gelatin	mPa-S at 10 sec⁻¹	On cooling	On heating	
2.0				0.5		25		Compatible
2.0				1.0		29	57	Compatible
2.0				2.0		~28	~57	Not compatible
2.0				4.0		—	—	Not compatible

Reprinted by permission.

Table 8
UNPLASTICIZED FILMS OF METHOCEL® CELLULOSE ETHER PRODUCTS

| | Typical data | |
Properties	METHOCEL® A	METHOCEL® E
Specific gravity	1.39	1.29
Area factor, in²/lb/mil	24,000	25,860
Moisture-vapor transmission, rate, 100°F, 90—100% R.H.	67.5 g/100 in²/24 hr/mil	65
Oxygen transmission rate, 75°F	25 cc/100 in²/24 hr/mil	70 cc/100 in²/24 hr/mil
Tensile strength, 75°F, 50% R.H.	8500—11,400 lb/in²	8500—8850
Elongation 75°F, 50% R.H.	10—15%	5—10%
Stability to ultraviolet (500 hr, Fadeometer exposure)	Excellent	Excellent
Resistance to oils and most solvents	Excellent	Excellent
Ultraviolet transmission 400 nm	55%	82%
(2 mil film) 290 nm	49%	34%
210 nm	26%	6%
Refractive index $N_D^{20°}$	1.49	
Softening point	—	240°C
Melting point	290—305°C	260°C
Charring temperature	290—305°C	270°C

Reprinted by permission of The Dow Chemical Company, 1978.

cations are described in Table 9. It can be seen from this table that it is the capability of these gums to contribute to fulfilling a variety of needs which makes them most valuable to the food industry. The most important applications and patent literature relating to these applications will be itemized.

A. Salad Dressings

Both methylcellulose and hydroxypropylmethylcellulose find applications in standard and low-calorie salad dressings.[39,40] They are effective stabilizers in French-type salad dressings and are recognized in the standard of identity for French-type salad dressings. In addition to stabilization, the methylcelluloses provide controlled pourability and mouthfeel characteristics through rheology adjustment. Dry-mix salad dressings also employ these gums commonly in conjunction with other stabilizers. The stability of the methylcelluloses to the presence of salt and sugar as well as the pH stability of these gums makes them especially valuable for salad dressing applications.

In low-oil and oil-free salad dressing applications, recent recommendations have been made for xanthan-methylcellulose blends.[39] These systems, employing about 1.1% total gum, are reported to have excellent viscosity stability at the resulting low pH levels and superior emulsion stabilization in the case of the 10% oil system. Bondi and Spitzer have patented an Italian-type low-calorie dressing using a methylcellulose and agar combination.[41] Spitzer and Nasarevich have patented a French dressing formulation employing methylcellulose in combination with agar and pectin.[42] Using a METHOCEL® A4M methylcellulose and xantham gum system, oil-free salad dressings with a caloric value of only 40 cal/100g have been successfully prepared.[43]

B. Baked Goods

A number of the functional properties of methylcellulose products are found beneficial in bakery applications. These include the gelation and binding characteristics of the gum

Table 9

FOOD INDUSTRY APPLICATIONS OF METHYLCELLULOSE AND HYDROXYPROPYLMETHYLCELLULOSE

Applications	Adhesion	Emulsion stabilization	Film/barrier	Freeze-thaw stability	Thermogelation	Hydration rate	Low pH stability	Moisture retention	Surfactancy	Suspending	Viscosity
Extruded foods	X	X			X			X	X		
Bakery goods	X	X			X			X	X		
Beverage — dry			X	X		X	X				X
Beverage — liquid		X	X			X	X			X	X
Desserts		X		X		X		X	X		X
Salad dressings (dry)		X				X	X		X	X	X
Salad dressings (liquid and spooned)		X					X		X	X	X
Fried foods	X		X	X	X			X			X
Pet foods	X					X		X	X		X
Sauces	X	X			X		X		X	X	X
Whipped toppings		X		X					X		X

which are similar to gluten, but are relatively unaffected by mineral salts, acids, and proteases. The thermogelation of the cellulosic helps promote gas retention during baking without increasing toughness in the finished product. Apparently the dough is strengthened by thermogelation at baking temperatures, resulting in finer structure and better yield. Both in yeast-leavened and chemically leavened systems, increased volumetric yield resulted from the incorporation of 0.2 to 0.4% METHOCEL® F4M.[44] In addition, methylhydroxypropylcellulose is used to aid in the production of low-protein, low-salt bread of improved palatability for consumption by persons suffering from celiac disease, phenylketonuriac disorders, uremia, or certain kidney disorders.[45] Elimination of protein containing ingredients in bread normally leads to a close-grained, nonresilient, dry, dense loaf. Incorporation of methylcellulose results in a loaf very similar to standard bread. Production of another alternate bread, a reduced calorie high-fiber diet bread, has been reported using hydroxypropylmethylcellulose flour.[46] Elimination of fat and reduction of the flour and sugar content in this bread formula led to a reduction in caloric content of 40%.

The surfactancy of methylcellulose products improves the whippability of cake batters in which a portion of the egg whites have been replaced by the methylcellulose product. Up to one half of the egg whites may be replaced on the basis of 0.5 parts of hydroxypropylmethylcellulose to replace 1 part of the egg white solids.[47] The batter foam stabilization resulting from the surface active behavior of the methylcellulose is further reinforced by the thermal gelation of the gum while baking. Reversion of the gel to the liquid state on cooling results in improved texture without any accompanying toughness.

The moisture retention functionality of the methylcellulose gums results in extended freshness. Taste panel tests conducted by the Dow Chemical Company indicated that cakes and raised doughnuts modified with METHOCEL® retained sweetness and moisture content longer than control samples.[48] Retrogradation of starch and staling were inhibited. Compressibility tests verified the texture conclusions of the taste panels.

Reduced fat absorption in raised doughnuts and cake doughnuts is attributed to the capability of the methylcellulose products to form oil insoluble films through the thermogelation process during frying of the doughnut. Compression tests indicated, as in the case of cakes, that a more tender product resulted from the inclusion of hydroxypropylmethylcellulose in the doughnut batter. It was reported as well that the batter was more tolerant to mixing conditions, undermixing or overmixing of batters not resulting in loss of quality.

A further application of the thermogelation property of methylcelluloses is the inclusion in fruit pie fillings to reduce boil over during the baking process.[49] The formation of the thermal gel inhibits this tendency while baking, yet does not result in a rigid filling when the pie is allowed to cool. Best economical results are obtained with mixtures of methylcellulose and starch to achieve desired body and mouthfeel characteristics at table conditions with boil over prevention at high temperatures.

C. Fried Foods

Methylcellulose and hydroxypropylmethylcellulose find use in batters for breaded foods and as a barrier coating in French frying of potatoes. A number of properties of the cellulosic are important in these applications including the adhesive, film-forming, thermal gelling, and noncharring characteristics. In breaded foods, hydroxypropylmethylcellulose solutions are used as replacements for egg-milk base batters for a number of reasons. Phase separation and viscosity drift are reduced in storage of batters prepared from hydroxypropylmethylcellulose as compared to egg-milk batters. Since the methylcellulose products are physiologically inert, they do not increase the caloric content of the food. They are also nonallergenic. The film-forming property of the methylcellulose prevents the absorption of oil into the interior of the breaded food, and, at the same time, helps to retain the natural moisture and juiciness of the food. In fact, hydroxypropylmethylcellulose has been reported to improve texture and moisture in rehydrated freeze-dried meats.[50] In this same patent, it is also described to perform as a coating to prevent freezer burn and promote moisture retention in frozen meats. The thermal gelation at high temperatures combined with the natural tackiness of the gum solution at low temperatures results in good pickup and adhesion of the breading material and superior retention of the breading material to the product when it is heated. In tests carried out evaluating batters containing METHOCEL® F50 and METHOCEL® K100 for breading a variety of food products including seafoods, meat patties, chicken and pork chops, the gums performed satisfactorily as adhesive agents. They were as effective as or better than egg-milk batters, allowed control of oil penetration into the food, imparted no flavor to the food, and showed equivalent or improved color. Potential for cost savings was considerable as the batter cost was only a fraction of the cost of egg-milk batter. The batter containing METHOCEL® was more stable to spoilage and the batter caused reduced frying oil consumption. In some cases, the flavor of the egg-milk batter is desirable in the food product. Methylcellulose may be used in these situations to effect cost savings by replacing only a portion of the batter solids with methylcellulose.

Other applications of methylcellulose products in the fried food areas have been in the areas of improved food patties and French fried potato products. Rivoche[51,52] found that incorporation of methylcellulose in patties of ground seafood, meat, vegetables, or mixtures of these resulted in a product which retained moisture and juices better on frying, and did not lose its shape or fall apart during cooking. The presence of the methylcellulose prevented the food patty from being ''blown apart'' by the flashing off of moisture and improved the texture of frozen food patties. Backinger and Meggison[53] mixed methylcellulose with cooked, mashed potatoes or with dehydrated potato flakes followed by rehydration. These mixtures were extruded and deep-fat fried. This procedure permits the use of small potatoes, slivers, and nubbins, in the manufacture of a French fried potato product substantially increasing yields and reducing costs. Willard and Roberts,[54] in describing the production of French fried vegetables, reported a synergistic effect of METHOCEL® products on the rehydration of dehydrated starchy vegetables. Gold[55] found that improved French fries resulted from dipping blanched and dewatered potato pieces in a solution of methylcellulose or hydroxypropylmethylcellulose prior to deep-fat frying. The hydrocolloid treatment is reported to yield improved color, texture, plate life, and a reduction in oil absorption. The reduced oil absorption results in a less greasy product and improved cooking economy from reduced oil losses. Keller[56] claimed an improved French fried potato mix by the use of a combination of a polygalactomannan gum and hydroxypropylmethylcellulose in the rehydration, binding, and extrusion of dehydrated mashed potatoes to yield a French fried potato product. The inclusion of a high amylose starch with a water-dispersible edible vegetable gum is reported to provide control of oil absorption and improved French fried product eating qualities in a rehydrated potato dough.[57]

D. Frozen Foods

The application of methylcellulose and hydroxypropylmethylcellulose to the processing of frozen foods is generally based on the capability of the hydrocolloid to retain moisture at subzero temperatures and to modify ice crystal formation in a variety of applications to yield freeze-thaw stability. The presence of methylcellulose in frozen pie fillings reduces water migration from the filling to the crust and thus prevents development of sogginess.[34,39] In reconstituted meat, fish, fowl, and vegetable products,[51,52] methylcellulose ethers function as a binder at low temperatures and help prevent freezer burn through improved moisture retention. Methylcellulose at a level of 0.3 to 3% in combination with a calcium salt of alginic acid is used to form stable gels of edible protein. The protein combined with moisture and gum binder may be extruded, shaped and frozen.[58] The inclusion of the cellulosic is critical to inhibition of syneresis in the product.

The incorporation of 0.2 to 0.4% METHOCEL® F400 methylcellulose in an imitation sour cream allowed freezing and subsequent thawing of the product while retaining the smooth texture of the original product.[59] Rivoche found additional application of methylcellulose in the preservation of foods, in particular fruits, but also vegetables, meats, and fish, by dehydration and freezing.[60] Rivoche also used methylcellulose in conjunction with starch in preparing frozen fruit pies. More attractive fruit fillings, stable gels at high as well as low temperatures, no leakage of juice and reduced tendency of juice to soak into the crust during baking were claimed.[61]

E. Whipped Toppings

The surface-active properties of methylcellulose and hydroxypropylmethylcellulose promote the use of these gums in the stabilization of nondairy whipped toppings for salads and desserts.[62] In this application, these products appear to be quite unique in their stabilizing ability and are normally used in combination with sodium alginate and additional emulsifiers. The emulsion obtained may be stored and used in a number of ways. It may be aerated either mechanically or by a gas such as nitrous oxide. The emulsion may be frozen either before or after aeration prolonging storage life, or it may be stored for extended times after heat sterilization.

It is also possible to dry the emulsion to produce a powder or to remove only a portion of the water to make a concentrate. Nesmick and Tatter[63] reported that methylcellulose ethers were satisfactory stabilizers in preparation of whippable compositions which could be spray-dried to obtain powders which could be stored for long periods without loss or change of flavor. Recombining the powder with water or milk while whipping results in a whipped composition suitable as a topping, filling, or component of custards or puddings, among other uses.

F. Beverages

There are a number of applications for methylcellulose in the beverage industry, including carbonated drink formulations, low-calorie soft drinks, and reconstituted dehydrated juices. Eagon and Greminger[64] used hydroxypropylmethylcellulose in combination with propylene glycol to prepare edible oil emulsions. Krumel et al. reported improvements on this work using lower molecular weight hydroxypropylmethylcelluloses at levels of 0.2 to 1.0 lb/gal of flavor oil emulsion to stabilize the emulsion used in carbonated soft drinks, particularly for cloudy beverages. Bakan reported[65] the encapsulation of oils, including lemon oil, using METHOCEL® followed by separation of the encapsulated oils and drying to obtain a dry encapsulated oil system.

Brenner[66] used methylcellulose in manufacture of sugar-free carbonated and noncarbonated beverages. The incorporation of 0.05 to 0.12% methylcellulose with added pectin provided body and texture to the drink to duplicate that which had resulted from the inclusion of sugar.

The application of methylcellulose products to the production of dehydrated juices is well known. Eddy patented the use of methylcellulose at levels of 0.1 to 4.0% of the fruit solids as a drying aid in the production of spray-dried fruit and vegetable juices.[67] The thermal gelation property of methylcellulose combined with its surfactant properties provides a stable foam which, when heated, does not break down, providing for ease of foam-mat drying.

A further application of methylcellulose has been as a suspending and a bodying agent in beverages. Love[68] reported that hydroxypropylmethylcellulose was preferred in a spray-dried skim milk chocolate drink. Farkas and Glicksman[69] have reported that a mixture of carboxymethylcellulose and methylcellulose provided a controlled rheology milk shake with a nearly constant viscosity over a broad temperature range.

G. Extruded Foods and Meat Analogs

The growing development of extruded foods and of meat analogs is expected to provide new applications for the methylcellulose gums. Past uses in such applications have included use as a binder in potato flakes for preparation of French fries, and as a binder, extrusion aid, and freeze-thaw aid in shaped fish and shellfish products prepared from minced fish and shellfish. Hydroxypropylmethylcellulose is used as a binder and extrusion aid in semi-moist pet foods where it also imparts improved sheen to the product.

H. Dietetic Foods

The methylcelluloses have a long history of application in reduced calorie foods and in specialized low-protein or low-gluten baked goods applications. These uses date back to the 1940s when Bauer and Wasson[70] proposed the use of methylcellulose gums in a number of dietetic food products for diabetics. These include syrups and various beverages which have improved organoleptic properties resulting from the inclusion of the methylcellulose gum. The tendency for syneresis in artificially sweetened jams and jellies is overcome by the use of methylcellulose.[71] Hydroxypropylmethylcellulose has recently been approved for use in confectionary applications.[13] Weight-reducing diets may include crackers containing methylcellulose to provide a sense of fullness and serve as a bulking agent to retain moisture and volume of food as it passes through the digestive tract.

Other dietetic applications which have been mentioned earlier in this chapter include the uses of methylcellulose in low-gluten and low-protein bakery products, low-oil and oil-free salad dressings, and low-calorie soft drinks.

I. Miscellaneous Uses

A number of more or less specialized uses, not fitting a specific category, have appeared in the literature.

Cyr[72] reported improvements in potato flakes obtained by use of methylcellulose as an emulsion stabilizer in the drying of a mixture of cooked mashed potatoes and an emulsifier.

Lantham and Seeley[73] patented a system of cellulosic gums used at a level of 3 to 5% in an egg composition. Hydroxypropylmethylcellulose was a critical component in this omelet mixture. It is claimed that the omelet composition will rise, not shrink when cooled or stored, and in fact, may be frozen without shrinkage. This system was reported to provide a precooked frozen omelet which upon reheating, had attractive appearance, flavor, and mouthfeel characteristics.

Glicksman[74] patented certain nonaqueous gelled compositions including those prepared from 0.1 to 10% of a methylcellulose or hydroxypropylmethylcellulose ether and propylene glycol, 1,3-butylene glycol or mixtures of these glycols.[74]

Lindblad[75] patented a food supplement composition containing 8 to 13% of methylcellulose. The intention of this lightweight food preparation was for use as an emergency ration. The purpose of the gum was to provide satisfactory organoleptic properties, prevent caking, and provide bulk.

Eppell[76] patented the use of methylcellulose with low methoxyl pectinate as an edible food bar coating. The purpose here was to impart a strong protective and edible coating to a fragile food bar, such as a cereal bar or a nut bar, and to improve handling, barrier, storage, and shipping properties.

In separate patents, Deacon et al.[77] and Chiu et al.[78] have patented the use of methylcelluloses in food casings. Deacon determined that the inclusion of a thermally gelling methylcellulose in a collagen coating at a preferred level of 5 to 20% prevents the disintegration of the coating while cooking the food item. This is of particular application in the preparation of collagen-cased sausages.

Chiu found that the inclusion of methylcellulose in the interior coating of a regenerated cellulose tubing used as a food casing for meats improved the ease of peeling the casing. Reduction of adhesion of the meat to the casing resulted in better appearance and less loss of the meat to the discarded casing.

REFERENCES

1. **Suida, W.,** Uber den Einflus der aktiven Atomgruppen in den textilefasern auf das Zustandekommen von Farburgen, *Monatsch. Chem.,* 26, 413, 1905.
2. **Lilienfield, L.,** A Process for the Manufacture of Ethers of Cellulose, Its Conversion Products and Derivatives, British Patent 12,854, 1912.
3. **Lilienfield, L.,** Cellulose Ether and Process of Making Same, U.S. Patent 683,831, 1928.
4. **Chowdhury, J. K.,** Uber Aether von Polysacchariden mit Oxysauren, *Biochem. Z.,* 148, 76, 1924.
5. **Savage, A. B.,** Water-Soluble Thermoplastic Cellulose Ethers, U.S. Patent 2,831,852, 1958.
6. **Savage, A. B.,** Process for Preparing Alkyl Hydroxyalkyl Cellulose Ethers and the Products Obtained Thereby, U.S. Patent 2,949,452,1960.
7. **Glicksman, M.,** *Gum Technology in the Food Industry,* Academic Press, New York, 1969, 437.
8. **Graham, H. E., Ed.,** *Food Colloids,* AVI, Westport, Conn., 1977.
9. **Krumel, K. L. and Sarkar, N.,** Flow properties of gums useful to the food industry, *Food Technol.,* 29, 36, 1975.
10. **Sarkar, N.,** Thermal gelation properties of methyl and hydroxypropylmethylcellulose, *J. Appl. Polym. Sci.,* 24, 1073, 1979.
11. **Savage, A. B.,** Cellulose ethers, in *Encyclopedia of Polymer Science and Technology,* Vol. 3, Mark, H. F., Gaylord, N. G., and Bikales, N. M., Eds., Interscience, New York, 1965, 459.
12. **Greminger, G. K. and Savage, A. B.,** Methylcellulose and its derivatives, in *Industrial Gums,* 2nd ed., Whistler, R. L., Ed., Academic Press, New York, 1973, chap. 18.
13. **Anon.,** Food additives permitted for direct addition to food for human consumption; hydroxypropylmethylcellulose, *Fed. Regis.,* 38273, 1982.
14. *United States Pharmacopeia,* 20th revision, July 1980.
15. Code of Federal Regulations, Title 21, Part 172, 1980; 172, 874 and 182.1480, 1977.
16. The Dow Chemical Company, Selecting METHOCEL® Cellulose Ethers, 1978.
17. **Anon.,** Methylcellulose in standards of identity for french and salad dressings, *Fed. Regis.,* 6711, 1959.
18. **Anon.,** Hydroxypropylmethylcellulose, Section 121.1021, Food Additives, *Fed. Regis.,* 8948, 1960.
19. **Anon.,** Hydroxypropylmethylcellulose in standards of identity in frozen desserts, *Fed. Regis.,* 18123, 1964.
20. **Nissan, A. H., Hunger, G. K., and Sternstein, S. S.,** Cellulose, in *Encyclopedia of Polymer Science and Technology,* Vol. 3, Mark, H. F., Ed., Interscience, New York, 1965, 131.
21. **Croon, I.,** The distribution of substituents in cellulose ethers, *Svensk. Papperstidn.,* 63, 247, 1960.
22. **McKelvey, J. B., Webre, B. G., and Klein, E.,** Reaction of epoxides with cotton cellulose in the presence of sodium hydroxide, *Tex. Res. J.,* 29, 918, 1959.
23. **Ramnas, O. and Samuelson, O.,** The rate of hydroxyethylation of cellulose in sodium hydroxide solution, *Svensk. Papperstidn.,* 76, 569, 1973.
24. **Ramnas, O.,** Determination of the ethylation rate of cellulose, *Acta Chem. Scand.,* 27, 3139, 1973.
25. British Celanese Limited, Celacol Courlose Water Soluble Cellulose Ethers.
26. **Waterhouse, B. R. and Higouchi, J.,** Marketing research report on cellulose ethers, in *Chemical Economics Handbook,* SRI International, in press.
27. **Glomski, R. L., Davis, L. E., and Grover, J. A.,** Water-soluble hydroxyethylmethylcellulose ether for latex paint, U.S. Patent 3,709,876, 1973.
28. The Dow Chemical Company, METHOCEL® Cellulose ethers, 1978.
29. **Muller, T. E. and Alexander, W. J.,** Characterization of the chain length distribution of wood cellulose by gel permeation chromatography, *J. Polym. Sci. Part C,* 21, 289, 1968.
30. **Alexander, W. J. and Muller, T. E.,** Evaluation of pulps, Rayon fibers, and cellulose acetate by GPC and other fractionation methods, *Sep. Sci.,* 6(1), 47, 1971.
31. **Wirick, M. G. and Elliot, J. H.,** A one-point intrinsic viscosity method for hydroxyethylcellulose, hydroxypropylcellulose, and sodium carboxymethylcellulose, *J. Appl. Polym. Sci.,* 17, 2867, 1973.
32. **Gibbons, G. C.,** *J. Text. Inst.,* 43, T25, 1952.
33. **Ganz, A. J.,** Cellulose hydrocolloids, in *Food Colloids,* Graham, H. E., Ed., AVI, Westport, Conn., 1977.
34. **Lindsey, T. A.,** Thermal gelling cellulose ethers for the food industry, presented at 33rd Annual Meeting of the Institute of Food Technologists, Miami Beach, June 10 to 13, 1973.

35. **Anon.,** Methylcellulose improves moisture retention, binding ability, *Baking Ind.,* 142, 10, 1975.
36. **Sarkar, N. and Greminger, G. K., Jr.,** Methylcellulose polymers — multifunctional ceramic processing aids, presented at The American Ceramic Society Meeting, Williamsburg, Va., September 16 to 19, 1979.
37. **Krumel, K. L. and Linsey, T. A.,** Nonionic cellulose ethers, *Food Technol.,* 30(4), 36, 1976.
38. **Krumel, K. L., Krasnoff, T. L., and Fiero, T. H.,** Edible oil emulsion, U.S. Patent 4,084,012, 1978.
39. The Dow Chemical Company, METHOCEL® in low-calorie salad dressing, 1974.
40. The Dow Chemical Company, METHOCEL® in food dressing, 1974.
41. **Bondi, H. W. and Spitzer, J. G.,** Low-Calorie Italian Dressing, U.S. Patent 2,916,384, 1959.
42. **Noserevich, L. S. and Spitzer, J. G.,** Low-Calorie French Dressing, U.S. Patent 2,916,383, 1959.
43. The Dow Chemical Company, Accelerated stability study of zero-oil salad dressings, 1978.
44. **Weaver, M. A., Bacon, K. D., and Greminger, G. K., Jr.,** Fried Cake Mix Containing Water-Soluble Cellulose Ethers, U.S. Patent 2,802,740, 1957.
45. **Wernecke, D. A.,** Dietetic Low Protein Bread Mix and Bread Produced Therefrom, U.S. Patent 3,567,461, 1971.
46. **Thompson, J. B.,** Low-Calorie Diet Bread, U.S. Patent 4,109,018, 1978.
47. **Weaver, M. A. and Greminger, G. K., Jr.,** Cake Mixes Containing Water-Soluble Cellulose Ethers, U.S. Patent 2,802,741, 1957.
48. The Dow Chemical Company, METHOCEL® in bakery goods, 1974.
49. The Dow Chemical Company, METHOCEL® food gums in bakery products, 1980.
50. **Harris, N. E. and Lee, F. H.,** Coating Composition for Foods and Method of Improving Texture of Cooked Foods, U.S. Patent 3,794,742, 1974.
51. **Rivoche, E. J.,** Frozen Food Patties and Method of Preparing Same, U.S. Patent 2,798,814, 1957.
52. **Rivoche, E. J.,** Method of Preparing Food Products, U.S. Patent 2,877,382, 1959.
53. **Bakinger, G. T. and Meggison, D. L.,** Method of Making a French Fried Potato Product, U.S. Patent 3,085,020, 1963.
54. **Willard, M. J., Jr. and Roberts, G. P.,** Producing French Fried Vegetables from Starch-Containing Dehydrated Vegetables and a Cellulose Ether Binder, U.S. Patent 3,399,062, 1968.
55. **Gold, W. L.,** Hydrocolloid Surface Treatment to Yield French Fried Potato Products, U.S. Patent 3,424,591, 1969.
56. **Keller, H. M.,** French Fried Potato Mix Made From Dehydrated Mashed Potatoes, U.S. Patent 3,468,673, 1969.
57. **Gremer, C. W.,** Method for Producing French Fried Potatoes, U.S. Patent 3,987,210, 1976.
58. **Carpenter, R. P., Cowie, W. P., Heyes, A., and Sutton, A. H.,** Protein Product, U.S. Patent 3,891,776, 1975.
59. **Calavert, P. B.,** Freeze and Thaw Stable Imitation Sour Cream, U.S. Patent 3,729,322, 1973.
60. **Rivoche, E. J.,** Process for Treating Food Products, U.S. Patent 3,010,831, 1961.
61. **Rivoche, E. J.,** Preparation of Food Products, U.S. Patent 3,052,558, 1962.
62. **Diamond, H. W. and Powell, E. L.,** Salad and Dessert Topping, U.S. Patent 3,868,653, 1959.
63. **Nesmick, P. P. and Tatter, C. W.,** Spray-Dried Whippable Food Composition, U.S. Patent 3,628,968, 1971.
64. **Eagon, B. M. and Greminger, G. K., Jr.,** Edible Oil Emulsions, U.S. Patent 3,906,626, 1959.
65. **Bakan, J. A.,** Method of Making Microscopic Capsules, U.S. Patent 3,567,650, 1971.
66. **Brenner, G. M.,** Sugarless Beverage, U.S. Patent 2,691,591, 1951.
67. **Eddy, G. W.,** Process of Drying Fruit or Vegetable Materials Containing Added Methylcellulose, U.S. Patent 2,496,278, 1950.
68. **Love, D. H.,** Reconstitutes without solids separating, *Food Process.,* 20, 1, 1959.
69. **Farkas, E. H. and Glicksman, M.,** Hydrocolloid rheology in the formulation of convenience foods, *Food Technol.,* 21(4), 49, 1966.
70. **Bauer, C. W. and Wasson, L. A.,** Sugar-free and nonglycogenic preparations for use by diabetic persons, *J. Am. Pharm. Assoc. Pract. Pharm. Ed.,* 9, 296, 1948.
71. **Windover, F. E.,** Aklyl and hydroxyalkylcellulose derivatives, in *Water Soluble Resins,* Davidson, R. L. and Sittig, M., Eds., Rheinhold, New York, 1962, 52.
72. **Cyr, J. W.,** Manufacture of Potato Products, U.S. Patent 3,056,683, 1962.
73. **Lantham, S. D. and Seeley, R. D.,** Egg Composition, U.S. Patent, 3,769,404, 1973.
74. **Glicksman, M.,** Non-Aqueous Gelled Compositions, U.S. Patent 3,949,103, 1976.
75. **Lindblad, R. L.,** Food Supplement Composition, U.S. Patent 3,058,828, 1955.
76. **Eppell, N. S.,** Edible Foodstuff Coatings, U.S. Patent 2,703,286, 1955.
77. **Deacon, J. D., Ferrers, H., and Kindleysides, L.,** Method of Making a Collagen-Coated Sausage, U.S. Patent 3,767,821, 1973.
78. **Chiu, H. S., Voo, D., and Standard, J. J.,** Food Casing and Method of Preparing Same, U.S. Patent 3,898,348, 1975.

Plant Seed Gums

INTRODUCTION

Martin Glicksman

Plant seeds are a traditional and ancient source of gums. Most seeds contain starches as the principal reserve food stored for use by the embryonic plant, but many seeds contain other polysaccharide polymers with functional gum-like properties which have served as a useful source of commercial hydrocolloids. Gums have been extracted or isolated since antiquity from many well-known seed sources such as quince, psyllium, flaxseed, locust bean (or carob), guar, tara, and tamarind, among others.[1,2] For food applications, only a few of these gums are important. The family comprising the galactomannans are the most widely used and play an important role in food technology.

Of the galactomannans, only guar gum and locust bean gum are used extensively in food product production and have unique functional properties that make them extremely applicable. Tara gum, another member of the galactomannan family, also has useful properties, lying somewhere between those of guar and locust bean gum, but it has not been truly commercialized and has not been approved for food applications. It is included in this section, because it does have the potential of becoming a significant commercial gum in the future if its novel properties are recognized and exploited.

Tamarind seed gum is not a galactomannan and is not related to guar, locust bean, or tara gums. It is not approved for food use in the U.S. However, tamarind seed gum is a recognized article of commerce and has been used in food applications in India and other parts of the Far East. It also has some unique, novel, and useful properties which can be employed in food applications. It has the potential for commercialization and clearance in the future and has therefore been included in this section on plant seed gums.

Historically, the galactomannan polysaccharides or gums have a venerable ancestry with locust bean gum dating back to the ancient Egyptian dynasties (circa 3000 B.C.) when they were used in the processes for mummification. Hence they have been referred to humorously as "Pharaoh's polysaccharides".[4] Biblical accounts in both the Old and New Testament continue the saga of the use of locust beans as food into the Christian era, and subsequently over the centuries into modern times.[5,6]

Guar is a more recent adjunct, coming into prominence during the Second World War as a result of a search aimed at finding gum replacements for locust bean gum and other gums that became unavailable because of the war. Its commercial development soared after the war to the point where current usage of guar in the U.S. dwarfs that of locust bean gum by a factor of 15 or 20 times that of locust bean gum.

Tara gum, available from Peru and used in the past as a tanning material, never became a factor in the food industry and is not approved for food applications. However, it does have interesting functional properties midway between locust bean gum and guar, and thus it certainly has the potential for development in the future.[7]

The galactomannan gums are found in the endosperm of various leguminous plants where they may have taxonomic importance, i.e., different species have various levels of the polymer as well as different ratios of D-mannose to D-galactose. Structurally, the galactomannans are based on a (1-4)-linked β-D-mannan backbone solubilized by side-chain substitutions of (1-6)-linked α-D-galactosyl groups present to various degrees in different galactomannans (Figure 1). In general, it has been found that the solubility of the polymer in water increases as the ratio of mannose to galactose (M/G) decreases.

In summary, the galactomannans from different sources vary in (a) mannose/galactose ratio, (b) distribution pattern of galactose residues on the mannose backbone, (c) molecular weight, and (d) molecular weight distribution.[3]

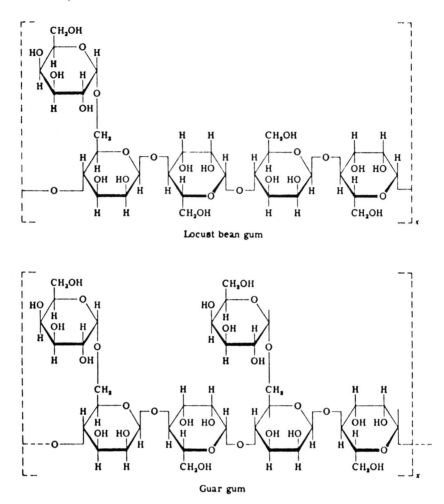

FIGURE 1.   Structure of locust bean gum and guar gum.

The mannose to galactose ratio of the three important galactomannans, varies approximately as follows:

	M/G
Locust (carob) bean gum	3.5 to 1
Tara gum	3 to 1
Guar gum	<2 to 1

REFERENCES

1. **Whistler, R. L.,** Polysaccharides, in *Encyclopedia of Polymer Science and Technology,* 11, 1969, 369—424.
2. **Cottrell, I. W. and Baird, J. K.,** Gums, in *Kirk-Othmer Encyclopedia of Chemical Technology,* Vol. 12, 3rd ed., 1980, 45—66.
3. **Robinson, G., Ross-Murphy, S. B., and Morris, E. R.,** Viscosity-molecular weight relationships, intrinsic chain flexibility, and dynamic solution properties of guar galactomannan, *Carbohydr. Res.,* 107, 17—32, 1982.

4. **Rees, D. A. I.,** *Biochem. J.,* 126, 257—273, 1972.
5. **Carlson, W. A.,** Carob as a food — a historic review, *Lebensm. Wiss. Technol.,* 13, 51—52, 1980.
6. **Rol, F.,** Locust bean gum, in *Industrial Gums,* 2nd ed., Whistler, R. L., Ed., Academic Press, New York, 1973.
7. **Anderson, E.,** Endosperm mucilage of legumes — occurrence and composition, *Ind. Eng. Chem.,* 41, 2887—2980, 1949.
8. **Dey, P. M.,** Biochemistry of plant galactomannans, *Adv. Carbohydr. Chem. Biochem.,* 35, 341—376, 1978.
9. **Dea, I. C. M., Morris, E. R., Rees, D. A., and Welsh, E. J.,** Associations of like and unlike polysaccharides: mechanism and specificity in galactomannans, interrelated bacterial polysaccharides, and related systems, *Carbohydr. Res.,* 57, 249—272, 1977.

Chapter 5

LOCUST/CAROB BEAN GUM

Carl T. Herald

TABLE OF CONTENTS

> And the same John had his raiment of camel's
> hair, and a leathern girdle about his loins; and his
> meat was locusts and wild honey.
> St. Matthew 3:4

I. HISTORICAL BACKGROUND

An ancient leguminous tree, *Ceratonia siliqua,* which grows in Mediterranean countries, bears a fruit pod commonly known as locust or carob bean. The tree has luxuriant perennial foliage, and after about 5 years begins to bear fruit, increasing slowly to maximum yield at 25 years when the tree is full grown (Figure 1). There is evidence that the carob fruit has been a useful food for centuries.

The history of the carob tree predates the Christian era by centuries. Bindings, treated with carob paste, were used by the ancient Egyptians to wrap mummies. Dioscorides, a Greek physician in the first century A.D., submitted that carob tree fruit was useful for syrupy extracts, a curative laxative, and had other therapeutic properties. Arabs utilized the carob kernels as weight standards to weigh gold, diamonds, and the like. The word *carat* is related to the botanical term *Ceratonia.*

In the earliest time of record, carob fruit was used as feed for horses, cattle, and pigs (hence the term ''swine's bread''). Carob pods have been consumed by poorer humans regularly.[1] Poor people in carob-growing areas continue to eat the pod pulp, which, because of a high sugar content (30 to 40%), is especially liked by children. During the Spanish Civil War, 1936 to 1939, people living in the carob-growing areas of southern Spain ate carob fruit and enjoyed better health than their Northern counterparts.

Some of the earliest references to use of carob as food are found in the Bible. Carob as food was cited during the preaching in the wilderness by John the Baptist. ''His meat was locusts and wild honey'' (Matthew 3:4) is believed to refer to wild carobs, and thus through the centuries carob bean has also been called St. John's bread, or Johannisbrot. The Prodigal Son in the Bible wasted his substance with riotous living and longed in vain to feast on the carob: ''And he would fain have filled his belly with the husks that the swine did eat: and no man gave unto him,'' (Luke 15:16).

Currently, the Biblical meaning of carob trees is commemorated in a traditional Jewish holiday. Jewish Arbor Day, TuBishvot, marks the tree-planting season of early Israel and generally falls in February. At this time, fruits distinctive to Israel are displayed and eaten. Carob, or *boksor* as called in Hebrew, is one of the fruits honored on this occasion.[2]

A. Agronomy

The carob tree thrives in the climate of Mediterranean countries. Producing areas are found in whole or parts of Spain, Italy, Greece, Cyprus, Algeria, Portugal, Morocco, Israel, and Turkey. The trees are used for shade in southern California, where the climate is similar to the Mediterranean sources.

Carob trees thrive in rocky soil unfit for most vegetation. Long roots penetrate the earth for depths of 60 to 90 ft, adapting the tree to areas of sparse rainfall. Carob fruits are pods of 4 to 8 in. in length and contain about 10 seeds or kernels per pod (Figure 2). Fructification begins in April or May, and by fall the pods are ripe and have turned from green to dark brown. A single carob tree may produce 1 ton of fruits or pods of which 8% are seed kernels.

B. Typical Composition

A tough, dark brown husk surrounds the carob bean kernel or seed. Efficient removal of the outer husk is essential to the production of a white finished powder. The locust or carob bean seed usually comprises 30 to 33% husk, 42 to 46% endosperm, and 23 to 25% germ.

After removal of the husk and germ, the endosperm is manipulated by selective milling to yield a range of finished products of varied particle size.

Typical analysis of locust bean gum, reported by Stein Hall and Co.,[3] is as follows:

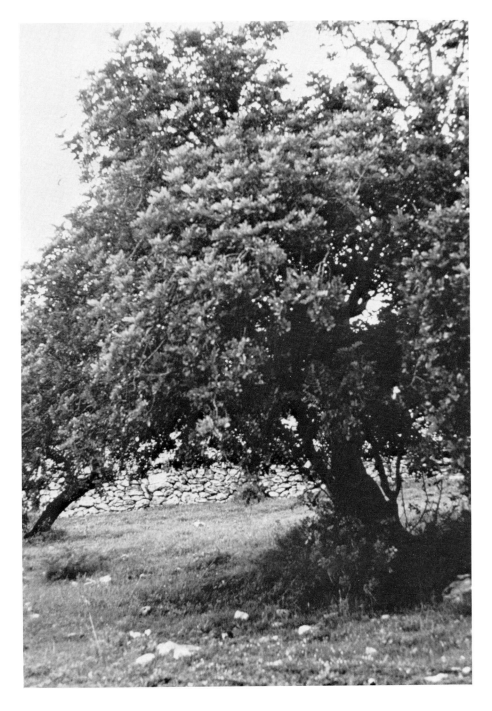

FIGURE 1. Locust or carob bean tree.

	Approximate values
Acid insoluble	0.5%
Protein	6.0%
Fat and alcohol soluble	0.5%
Ash	1.0%

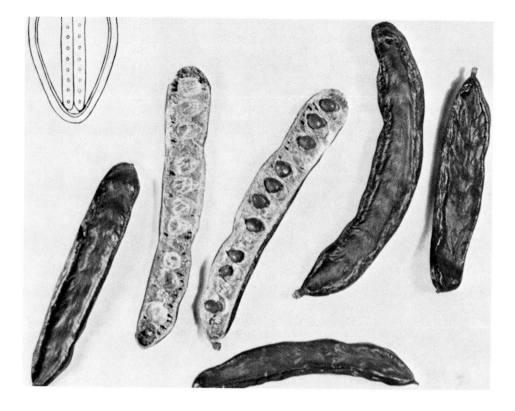

FIGURE 2.   Locust or carob pods.

Moisture	12.0%
Gum content	80.0%
Gum content, calculated on dry substance	88.0%
Starch	Negative
Test for reducing sugars after hydrolysis with HCl	Positive
Viscosity, 1% solution	Minimum 2000 cps

C. Trade Names

Assigned common names to the processed endosperm fraction of *Ceratonia siliqua* seeds are carob bean gum, locust bean gum, algaroba, and St. John's gum. The gum is marketed under a number of trade names throughout the world: Arobon, Gum Gatto, Gum Hevo, Gum Tragon, Jondagum, Johannisbrotmehel, Lakoegum, Lupogum, Luposol, Rubigum, Tragarab, and Tragasol.

II. DESCRIPTION

The description and specifications for locust bean gum as published in the Food Chemicals Codex III[4] are as follows:

A. Description

''A gum obtained from the ground endosperms of *Ceratonia siliqua* (L.) Taub. (Fam. *Leguminosae*). It consists chiefly of a high molecular weight hydrocolloidal polysaccharide, composed of galactose and mannose units combined through glycosidic linkages, which may be described chemically as a galactomannan. It is a white to yellowish white, nearly odorless powder. It is dispersible in either hot or cold water forming a sol, having a pH between 5.4

Table 1
MAXIMUM USAGE LEVELS
PERMITTED FOR LOCUST BEAN GUM[a]

Food (as served)	Percent
Baked goods and baking mixes	0.15
Beverages and beverage bases, nonalcoholic	0.25
Cheeses	0.80
Gelatins, puddings and fillings	0.75
Jams and jellies, commercial	0.75
All other food categories	0.50

[a] Title 21, Code of Federal Regulations, Parts 100 to 199, 184.1343.

and 7.0, which may be converted to a gel by the addition of small amounts of sodium borate.''

B. Specifications[4]

Galactomannans	Minimum of 73.0%
Acid-insoluble matter	Max. 5%
Arsenic (as As)	Max. 3 ppm (0.0003%)
Ash (Total)	Max. 1.2%
Heavy Metals (as Pb)	Max. 20 ppm (0.002%)
Lead	Max. 10 ppm (0.001%)
Loss on drying	Max. 15%
Protein	Max. 8%
Starch	Passes test

III. REGULATORY STATUS

A. Status

A summary[5] of available scientific literature from 1920 to 1972 related to the safety of carob bean gum as a food ingredient was prepared for the U.S. Food and Drug Administration. The report[5] summarizes chemical information, biological data, and biochemical aspects of carob bean gum and includes more than 100 appropriate references.

Locust (carob) bean gum as a direct human food ingredient is classified as generally recognized as safe (GRAS) by the Food and Drug Administration. Under a reaffirmation program, uses and specific maximum levels were published. The final order as published in the Code of Federal Regulations is given in Table 1.[6]

B. Labeling

The label requirement for locust (carob) bean gum is described in the Code of Federal Regulations, Title 21, 101.4. The regulation states that the food ingredient shall be declared by common or usual name in descending order of predominance. Exemption from labeling under special described circumstances is delineated under regulation 101.1.

IV. AVAILABILITY

In common with other seed gums, weather is an important yield determining factor for locust bean gum.[7] Hot dry weather is the ideal climate. Freezing reduces yield and may destroy the crop; cold weather and insufficient sun causes small fruit. Excessive rainfall encourages large pods with a small number of kernels.[8]

According to data supplied by the Bureau of Census, 7,231,000 lb of locust bean gum were imported into the U.S. in 1979. Estimated consumption of locust bean gum in 1967 was estimated at 3 million lb.[4]

Prices for locust bean gum rose sharply in the second half of 1979. Prices in the $2 to $3/lb range were reported in 1980.[9]

V. STRUCTURE

The locust bean gum molecule is a straight chain of D-mannopyranosyl units linked (1→4), with every fourth or fifth D-mannopyranosyl unit substituted on C-6 with a galactopyranosyl unit. Structurally, locust bean gum and guar gum have similarities with locust bean gum having a smaller number of D-galactose side groups[10] (Figure 1). Locust bean gum has a molecular weight of 310,000.[2,8]

VI. PROPERTIES

Locust bean gum is slightly soluble in room temperature water and must be heated to 75 to 85°C for complete hydration and viscosity development. A 1% locust bean gum solution hydrated at 25°C develops 100 cps viscosity.[11]

Obviously, a locust bean powder with a viscosity potential greater than 100 cps at room temperature indicates a product that may be contaminated with a faster hydrating colloid such as guar.

On heating a suspension of locust bean gum and water, viscosity decreases slightly followed by a sharp increase at about 45°C and on until about 85°C is reached. As the solution is cooled down to room temperature, a further viscosity increase is observed.

By thorough mixing of 4 parts locust bean gum with 6 parts corn sugar followed by wetting, heating, steaming, drying, and grinding, a cold-water soluble locust bean gum was claimed by Leo;[12] however, maximum viscosity development required heat. Almost identically shaped viscosity curves are realized from locust bean and guar gum.[13] A typical 1% solution of locust bean gum, on an "as is" moisture basis, has a viscosity range of 2400 to 3200 cps.[14,15] Variables such as granulation or particle size, moisture content, and galactomannan content will influence final viscosity.

The viscosity of locust bean gum solutions is not materially affected by salts commonly used in the food industry. Since locust bean gum is nonionic, the polymer is stable between pH 3.5 to 11.0.

Locust bean gum solutions tend to have opacity due to insoluble protein and cellulose impurities. Obviously, cloudiness and opacity is a disadvantage in finished products requiring sparkling clarity.

Texture modification of carrageenan water gels is achieved through the incorporation of locust bean gum.[16,17] Cohesive, brittle gels normal to carrageenans develop textures similar to gelatin gels. Locust bean-carrageenan gels gel above room temperature and maintain constant texture at room temperature and through storage. Combined locust bean gum and carrageenan have been used in a dry mix capable of forming edible gel structure. One such dry mix composition teaches the use of 86 parts sucrose, 1.36 parts carrageenan, 0.64 parts potassium chloride, 1.30 parts locust bean gum, 1.4 parts fumaric acid, and 1.6 parts sodium fumarate. The mix is dissolved with stirring in 1 pt of boiling water, and then the mixture sets into a gel.[18]

Synergistic viscosity results from the combination of locust bean and xanthan gums.[19] A thermally reversible gel is formed when locust bean gum and xanthan are heated to above 130°F and cooled.[20] Maximum gel strength or viscosity are reached at equivalent amounts of each hydrocolloid. Best gel strength or viscosity is reached at neutrality to pH 8, and a decided viscosity drop is evidenced at pH levels lower than 5. Final viscosity is also a function of time and shear input.

VII. APPLICATIONS

The broad application of locust bean or carob bean gum in foods is easily seen by reading ingredient labels in the grocery store. Major use areas are frozen desserts, cultured dairy products, cheese products, sauces, dips, and dressings.

A. Frozen Desserts
Locust bean gum has been used as a basic ice cream stabilizer for at least 50 years.[21] When used as a sole stabilizer for ice cream mix, locust bean gum causes a phase separation in ice cream mix.[22] The addition of carrageenan with locust bean gum results in homogeneous ice cream mix dispersions.

McKiernan[23] reported that locust bean gum met the ideal gum requirements for ice cream stabilizer for the following reasons:

1. It is readily dispersible in cold mix and requires no special handling or neutralization.
2. It is inert to both acidity (lactic acid) and calcium salts found in the mix.
3. It produces a uniform, medium, reproducible viscosity, which is an indication only of "bound water", i.e., it develops a basic viscosity that is not destroyed by agitation.
4. It does not cause mix separation when used with an Irish Moss (carrageenan) adjunct.
5. It cools uniformly and permits easy incorporation of air into the mix.
6. It provides exceptionally high water absorption, resulting in excellent body, smooth, fine texture, and chewiness to ice cream.
7. Its resultant viscosity-temperature curve does not fall rapidly, which results in superior heat shock resistance.
8. It does not contribute any taste or flavor-masking properties to the mix.
9. Pound for pound, it provides a higher degree of stabilization and heat shock resistance than that of any available gum.

Combinations of guar gum and locust bean gum are frequently used by compounders of proprietary ice cream stabilizers. Locust bean gum contributes a unique, short smoothness to ice cream, whereas guar contributes to body and chewiness.

Rapid freezing with agitation and the presence of a hydrophilic colloid promotes fine ice crystal structure in ice cream. Locust bean gum helps to maintain the small ice crystals during periods of temperature fluctuations.

A typical proprietary ice cream stabilizer label ingredient discloses: mono- and diglycerides, guar gum, locust bean gum, polysorbate 80, and calcium carrageenan. Locust bean gum usually accounts for 30 to 50% of the total gums in the above combination.

Locust bean gum is frequently a preferred colloid for sherbet stabilization. Locust bean gum imparts a desirable short, smooth texture to sherbet. Other gums such as guar or karaya may be used in combination with locust bean gum to achieve textured variations.

Locust bean gum has also been recommended as a stabilizing colloid in ices.[21]

B. Cultured Dairy Products
Cream cheese and similar products are a major outlet for locust bean gum. Besides controlling moisture in the end product, a unique texture effect is realized from the use of locust bean gum. Other gums such as guar have not been satisfactory alternatives. Despite the chemical similarity between guar and locust bean gums, the achieved technical result in cream cheese is quite different. Apparently, the milk casein polymers and locust bean gum are sterically aligned in such a way that new desirable complexes are formed.

Cream cheese is commonly manufactured by the hot pack process. In one method, moisture-adjusted curd is dumped into a steam-jacketed kettle and heated to 160 to 170°F. While heating and stirring, salt and about 0.5% locust bean gum are added. The heated mixture is then homogenized at about 2500 lb pressure, single stage, and packaged while hot.

Sour cream also benefits from the addition of locust bean gum. As with cream cheese, a desired texture effect is achieved in the end product and moisture is controlled. Carrageenan must be added with locust bean gum to prevent phase separation.

Dressing for cottage cheese curd is frequently stabilized with combinations of locust bean gum and carrageenan. Commercial cottage cheese is commonly two thirds cheese curd admixed with one third creaming mixture of 12% butterfat. Locust bean gum aids moisture control in the finished product and adds lubricity. Lubricity helps maintain curd integrity during processing and also aids eating.

That locust bean gum does not develop appreciable viscosity under pasteurization temperatures is an advantage in the preparation of large cottage cheese dressing batches. The proprietary mixture may be dispersed by high shear in one operation, whereas a cold-soluble gum may require two or three dispersion operations due to viscosity limitations. Synergistic mixtures of locust bean gum and xanthan gum are also used in cottage cheese dressings, dips, and dressings.

C. Cheese

The incorporation of locust bean gum in ricotta cheese operations is beneficial to manufacturing and quality. Coagulation is accelerated and curd yield is improved by as much as 10%. Whey leakage from the finished product is also controlled.[8]

Processed cheese and cheese spreads are improved by the use of locust bean gum. The colloid aids in controlling water and results in a firm, homogeneous, spreadable product. A concentration of about 0.6% gum is incorporated with the other ingredients prior to processing.[24]

The baking characteristics of spray-dried bakers cheese products made with acidified milk and cream combinations are enhanced by the incorporation of locust bean gum.[25]

D. Meat Products

In some countries, locust bean gum is used as an ingredient in specialty products such as salami, sausage, and bologna. Due to the lubricity imparted by the gum, extrusion and stuffing is facilitated. The water retention property of the gum reduces finished product weight loss during storage. Consequently, technical and economic advantages are achieved.[2]

Meat product analogs prepared from plant proteins have incorporated locust bean gum to provide essential meatlike textural chew.[26-28]

E. Bakery Products

Bread flour supplemented with locust bean gum produces doughs with constant properties and enhanced water-binding characteristics. Moreover, yields are improved and the baked products remain soft and palatable for a longer time.

The gum also benefits cake and biscuit doughs in that yields are improved, and the finished products have a longer shelf life. Cakes are more easily removed from the pan, and the texture is firmer, which in turn, facilitates slicing.

Combinations of modified starches and locust bean gum are extremely useful in stabilizing fruit fillings used in baked goods such as pie or Danish pastry. A typical Danish pastry filling contains 5% modified starch. This may be reduced to 4% starch and 0.5% locust bean gum for a product stable to boil out during baking, and also having improved fruit flavor.

Locust bean gum has been reported as a very satisfactory stabilizer for canned berry and berry-apple pie fillings.[29] Combinations of locust bean gum and starches for frozen pie fillings have been found to be functionally acceptable.[30]

F. Dessert Gels

Gels achieved through the interaction of locust bean gum and calcium carrageenan are of interest to the food industry. A fruit gel stable to storage temperatures above 70°F was prepared with the following formulation from Billerbeck:[33]

Ingredient	Formulation, lb/100 gal
Montmorency cherry juice concentrate	27
Dark sweet cherry concentrate	27
Granulated white sugar	177
Locust bean gum/calcium carrageenan (Gelcarin® DG), 1.5/1 weight ratio	4
Tripotassium citrate	2
Carrageenan (Gelcarin® HWG)	0.8
Ascorbic acid	0.7
Salt	0.5
Calcium sulfate	0.5

Syneresis-resistant dessert gels have been prepared from locust bean gum, potassium, and calcium-sensitive carrageenans.[34] Reportedly, long storage times at accelerated temperatures resulted in a minimum of syneresis. Obviously, such gels have great utility in countries lacking modern refrigeration.

Gel systems were designed to support chunks of fruit, including pineapple, and the hydrocolloid portion comprised low methoxy pectin, locust bean gum, potassium-sensitive carrageenan and furcellaran, and calcium-sensitive carrageenan.[35]

An edible gel dessert containing uniformly distributed fruit particles utilizing the locust bean gum and carrageenan combination was described by Klein and Cerchia.[36] The dessert mixture was heated to a sterilization temperature and then rapidly cooled to 140 to 155°F and then finally cooled to about 105°F while rotating the dessert containers longitudinally until a gel formed and all fruit pieces remained in constant position.

G. Miscellaneous

The pod of carob or locust bean gum continues to serve as an important source of flavor. Concentrated extracts of carob bean pod are sold as St. John's Bread extract by leading botanical precessors. Carob extract may be used alone or as a component of major flavors. From a flavor standpoint, carob extract is classified aromatic and sweet.[31] During periods of vanilla or cocoa shortages, carob extract is an important component of substitute compounded flavors. Carob extract is also useful as a smoking tobacco flavor.

So-called natural or health foods frequently incorporate carob flavor. Extravagant claims are made for these products from a health and medicinal standpoint. Recipes for carob replacement of chocolate in cakes, cookies, candies, ice cream, and malted milk have been published.[32] As a flavor, or part of a flavor, St. John's Bread extract is a useful tool; however therapeutic medicinal advantages await discovery.

Various researchers have found locust bean gum useful for the treatment of diarrhea, a body improver in reconstituted dry milk, an antioxidant for butterfat, a stabilizer in citrus juice products and in an aerated dessert.[2]

Locust bean gum continues to be used as a thickening and bodying agent in commercial barbecue sauce.

REFERENCES

1. **Moldenke, H. N.,** The economic plants of the Bible, *Econ. Bot.,* 8, 152, 1954.
2. **Glicksman, M.,** Locust bean gum, in *Gum Technology in the Food Industry,* Academic Press, New York, 1969, chap. 5.
3. Stein, Hall & Co., Inc., Locust Bean Gum, Product Bulletin, 4 pages.
4. *Food Chemicals Codex,* 3rd ed., National Academy Press, Washington, D.C., 1981, 174—175.
5. National Technical Information Service, GRAS (Generally Recognized as Safe) Food Ingredients — Carob Bean Gum (Locust Bean Gum) PB 221 203, Springfield, Va., 1974, 24.
6. Title 21, Code of Federal Regulations, Parts 100 to 199, 184.1343, 1979, 748.
7. **Meer, G., Meer, W.-A., and Tinker, J.,** Water-soluble gums, their past, present and future, *Food Technol. (Chicago),* p. 22, November 1975.
8. **Rol, F.,** Locust bean gum, in *Industrial Gums,* 2nd ed., Whistler, R. L., Ed., Academic Press, New York, 1973, chap. 15.
9. Chemical Marketing Reporter, March 10, 1980.
10. **Smith, F. and Montgomery, R.,** *The Chemistry of Plant Gums and Mucilages and Some Related Polysaccharides,* Reinhold, New York, 1959, 627.
11. **Goldstein, A. M., Alter, E. N., and Seaman, J. K.,** Guar gum, in *Industrial Gums,* 2nd ed., Whistler, R. L., Ed., Academic Press, New York, chap. 14.
12. **Leo, A. J.,** U.S. Patent 2,949,428, 1960.
13. **Smith, P.,** Limitless prospects seen for use of natural gums in candy, *Candy Ind. Confect. J.,* 124, 74, 1965.
14. Henkel, Food Ingredients Division, Super Col® Guar, Locust Bean Gums Bulletin, 11 pages.
15. Hercules Inc., Locust Bean Gum Bulletins, 959, 960, and 961, Wilmington, Del.
16. **Baker, G. L.,** U.S. Patent 2,466,146, 1949.
17. **Baker, G. L.,** U.S. Patent 2,669,519, 1954.
18. **Campbell, A. D.,** U.S. Patent 3,031,308, 1962.
19. **Rocks, J. K.,** Xanthan gum, *Food Technol. (Chicago),* 25(5), 22, 1971.
20. **Kovacs, P.,** Useful incompatibility of xanthum gum with galactomannons, *Food Technol. (Chicago),* 27(3), 26, 1973.
21. **Dahle, C. D. and Collins, W. F.,** Basic Stabilizers in the Ice Cream Industry, International Association Ice Cream Mfgers. Proc., 43rd Annual Convention, 1947, 50.
22. **Potter, F. E. and Williams, D. H.,** Stabilizers and emulsifiers in ice cream, *Milk Plant Mon.,* 39(4), 76, 1950.
23. **McKiernan, R. J.,** The Role of Gums in Stabilizers, paper presented at the Michigan Dairy Manufacturers Annual Conference, Michigan State University, East Lansing, 1957.
24. **Werbin, S. J.,** Practical Aspects of Viscosities of Natural Gums, Physical Functions of Hydrocolloids, American Chemical Society, 1960, 5.
25. **Roundy, Z. D. and Osmond, N. R. H.,** U.S. Patent 2,956,885, 1960.
26. **Anson, M. L.,** U.S. Patent 2,833,651, 1958.
27. **Anson, M. L. and Pader, M.,** U.S. Patent 2,830,902, 1958.
28. **Anson, M. L. and Pader, M.,** U.S. Patent 2,879,163, 1959.
29. **Moyls, A. W., Atkinson, F. E., Strachan, C. C., and Britton, D.,** Preparation of canned berry and berry-apple pie fillings, *Food Technol. (Chicago),* 9, 629, 1955.
30. **Carlin, G. T., Allsen, L. A., Becker, J. A., Logan, P. O., and Ruffley, J., Jr.,** Pies — How to Make, Bake, Fill, Freeze, and Serve, Tech. Bull. 121, National Restaurant Association, Food and Equipment Research Department, Chicago, 1954.
31. **Merory, J.,** *Food Flavorings, Composition, Manufacture, and Use,* AVI, Westport, Conn., 1968, 82.
32. **Binder, R. J., Coit, J. E., Williams, K. T., and Brekke, J. E.,** Carob varieties and composition, *Food Technol. (Chicago),* 13, 213, 1959.
33. **Billerbeck, F. W.,** U.S. Patent 3,367,783, 1968.
34. **Moirano, A. L.,** U.S. Patent 3,445,243, 1969.
35. **Moirano, A. L.,** U.S. Patent 3,556,810, 1971.
36. **Klein, R. A. and Cerchia, A.,** U.S. Patent 3,658,556, 1972.

Chapter 6

GUAR GUM

Carl T. Herald

TABLE OF CONTENTS

> And Moses stretched out his hand over the sea;
> and the Lord caused the sea to go back by a
> strong east wind all that night, and made the sea
> dry land, and the waters were divided.
> Exodus 14:21

I. HISTORICAL BACKGROUND

A. Agronomy

The guar plant, *Cyamposis tetragonolobus* of the *Leguminosae* family, has a history reaching far into the past of the subcontinent of Bangladesh-India-Pakistan. The legume is an important source of nutrition to animals and humans, regenerates soil nitrogen, and the endosperm of guar seed is an important hydrocolloid widely used across a broad spectrum of industries.

Ravel[1] writes that guar is known among the various Indian languages as follows: Gujaratic-Sindhi, Gavari-Marathi, Gavar Fali-Hindi, Sim-Bengali, Guvar-Urdu, Kottavari-Tamil, Amar Ppayara-Malayalam, derived from the Sanskrit work Gau-ahar. "Gau" means cow and "-ahar" means food. Ravel comments on the ancient unknown agronomist who in one stroke fed his beast of burden, promoted milk production, provided a human food, enriched soil through nitrogen fixation, and created a versatile hydrocolloid that is used in various forms in the food, dairy, cosmetic, pharmaceutical, oil, mining, paper, and other industries.

Guar was introduced into the U.S. in 1903 as a possible cover crop in the semiarid regions of Texas, Arizona, and California. A severe locust bean gum shortage shortly after the Second World War adversely affected the paper and textile industries. Guar gum was found to be the most suitable substitute for scarce locust bean gum.[2]

In 1953, the gum was produced in commercial quantities by Stein, Hall & Co. and General Mills. Subsequently, European producers of locust bean gum began to process guar gum.

The guar legume plant is extremely drought-resistant and thrives in semiarid regions where most plants perish. Guar requires reasonably warm weather and a growing season of 20 to 25 weeks. The plants are 3 to 4 ft high, vertically stalked, and lend themselves to modern combine harvesting operations (Figure 1). The seed pods are 2 to 3 in. long and each contains 6 to 9 pea-shaped seeds weighing about 3 g/100 seeds[3] (Figure 2). Seeds are commonly 2 to 4 mm in diameter.

Modern agronomy programs are aimed at new guar varieties with superior characteristics such as higher yields and shorter growing seasons. Moreover, guar agronomy programs are under way in South America, Australia, Africa, and other areas to identify alternative crop sources.[4] The major current sources of guar are Pakistan and India. Some guar is grown in the arid regions of Texas, the acreage dependent upon economic return from competitive crops such as cotton.

B. Typical Composition

The mechanically disintegrated or ground endosperm of the guar seed constitutes the hydrocolloid portion commonly known as guar gum. Specialized mechanical treatment removes the husk and seed germ from the endosperm portion. A typical guar seed comprises

FIGURE 1. Field of guar.

hull, 14 to 17%; germ, 42 to 47%; and endosperm, 35 to 40%. A portion of endosperm is lost in processing. A typical analysis of a commercial guar gum marketed by Stein, Hall & Co.[5] is as follows:

Galactomannan	78—82%
Water	10—13%
Protein	4—5%
Crude fiber	1.5—2.0%
Ash	0.5—0.9%
Ether extractables	0.5—.075%
Iron	trace
Heavy metals	0
Arsenic	0

C. Trade Names

Significant amounts of this gum are marketed under its common name — guar gum. Trade names for food grade guar include Jaguar® and Supercol®. Jaguar® is the registered trademark of Stein, Hall & Co., Inc. (Celanese, Inc.) and Supercol® is the registered trademark of General Mills, Inc., now belonging to the Henkel Corp.

II. DESCRIPTION

A. Description

The Food Chemicals Codex[6] describes guar gum as follows:

FIGURE 2.   Guar pods and seeds.

Table 1

MAXIMUM USAGE LEVELS PERMITTED FOR GUAR GUM[9]

Food (as served)	Percent	Function
Baked goods and baking mixes	0.35	Emulsifier and emulsifier salts, formulation aid, stabilizer, and thickener
Breakfast cereals	1.2	Formulation aid, stabilizer, and thickener
Cheese	0.8	Formulation aid, stabilizer, and thickener
Dairy products analogs	1.0	Firming agent, formulation aid, stabilizer and thickener
Fats and oils	2.0	Firming agent, formulation aid, stabilizer, and thickener
Gravies and sauces	1.2	Formulation aid, stabilizer, and thickener
Jams and jellies, commercial	1.0	Formulation aid, stabilizer, and thickener
Milk products	0.6	Formulation aid, stabilizer, and thickener
Processed vegetables and vegetable juices	2.0	Formulation aid, stabilizer and thickener
Soups and soup mixes	0.8	Formulation aid, stabilizer, and thickener
Sweet sauces, toppings, and syrups	1.0	Formulation aid, stabilizer, and thickener
All other food categories	0.5	Emulsifier and emulsifier salts, firming agent, formulation aid, stabilizer and thickener

"A gum obtained from the ground endosperms of *Cyamopsis tetragonolobus* (L.) Taub. (Fam. *Leguminosae*). It consists chiefly of a high molecular weight hydrocolloidal polysaccharide, composed of galactan and mannan units combined through glycosidic linkages, which may be described chemically as a galactomannan. It is a white to yellowish white, nearly odorless powder. It is dispersible in either hot or cold water forming a sol, having a pH between 5.4 and 6.4, which may be converted to a gel by the addition of small amounts of sodium borate."

B. Specifications

Food grade guar gum specifications are defined in the Food Chemicals Codex[6] as follows:

Galactomannans	Minimum of 66.0%
Acid-insoluble matter	Max. of 7%
Arsenic (as As)	Max. of 3 ppm (0.0003%)
Ash (total)	Max. of 15%
Heavy Metals (as Pb)	Max. of 20 ppm (0.002%)
Lead	Max. of 10 ppm (0.001%)
Protein	Max. of 10%
Loss on drying	Max. of 15%
Starch	Passes iodine color test

III. REGULATORY STATUS

A. FDA Status

A summary report of available scientific literature from 1920 to 1972 related to the safety of guar gum as a food ingredient was prepared for the U.S. Food and Drug Administration. Chemical information, biological data, and biochemical aspects of guar gum are outlined in a 35-page summary with 116 references by the National Technical Information Service.[7]

As a direct human food ingredient, guar gum is classified as generally recognized as safe (GRAS) by the Food and Drug Administration. The proposed affirmation of GRAS status with specific limitations as a direct human food ingredient and affirmation of GRAS status as an indirect human food ingredient was published in the Federal Register.[8] The final order as published in the Code of Federal Regulations[9] is summarized in Table 1.

B. Labeling

The labeling requirement for guar is described in the Code of Federal Regulations, Title 21, 101.4. The regulation states that the food ingredient shall be listed by common or usual name in descending order of predominance. Exemption from labeling under special described circumstances is delineated under regulation 101.1.

IV. MANUFACTURE

Guar gum may be supplied by basic manufacturers or importers. Some principal suppliers are Celanese Plastics and Specialties Co. (formerly Stein, Hall & Co.); Henkel Corporation (formerly General Mills) and Hercules, Inc. (formerly Cesalpinia SPA).

A. Processing

Processing of guar seeds follows the basic procedure of grain or seed technology. Special mechanical treatment removes the husk and seed germ from the endosperm portion. The intact endosperm portion is commonly called a split. Guar gum is a desirable animal feed; hence, India and Pakistan do not permit the exportation of seed. Splits are frequently macerated in the source country into the finished product; however, Celanese Plastics and Specialties Co. imports splits and completes processing in the U.S. Guar endosperm raw material costs are intimately related to by-product meal markets.

B. Availability

A salient feature of the cultivated guar plant is the renewable resource feature. Crops are grown annually with a growing season of 20 to 25 weeks. Although the plant is drought resistant, weather conditions are important.

Guar may be plowed down for its soil-building properties as green manure.

Cattle consume half of the guar seed grown, with the remainder processed for industrial, edible, and reseeding purposes. Some estimate that 200 to 300 million lb of guar gum is used on a world-wide basis. In the U.S., about 104 million lb are imported, of which 15 to 25 million lb are believed used for edible purposes. Industrial uses are by far the largest and have been reviewed by Goldstein et al.[10]

An annual growth rate of 15% is believed. The strong growth rate reflects activity in the use of guar gum as a functional aid in secondary oil recovery, and as a flocculant in the mining industry. Guar prices have recently been in the $0.75 to $0.80/lb range.[11]

V. STRUCTURE

The guar molecule is a straight chain galactomannan with galactose on every other mannose unit. Beta 1→4 glycosidic linkages couple the mannose units and the galactose side drains are linked through alpha (1→6) as seen in Figure 1 of the Introduction. The molecular weight of guar has been reported as 220,000.[12]

Comparison of guar and cellulose molecular structure reveals that each is a rigid rod-like polymer because of the beta linkage between the monomer units. The spatial position of the hydroxyl groups differs between cellulosics and galactomannans. The guar hydroxyls are in the *cis* (same side) position and cellulosic hydroxyls are in the *trans* position. The *cis* position is important since adjacent hydroxyl groups reinforce each other in hydrogen-bonding reactions. Consequently, guar forms viscous aqueous dispersions and hydrogen bonds with hydrated cellulose. Alpha-cellulose is not soluble in water.

VI. PROPERTIES

A. Dispersion

Guar gum forms a colloidal dispersion (hereinafter referred to as guar solutions) to yield a highly viscous system. Guar solutions may be prepared in the laboratory or the manufacturing plant, by a variety of available techniques:

1.   Fairly rapidly sprinkle the gum into a vortex of highly agitated water. Cold water aids dispersion and slows hydration rate. Slow addition of the gum permits hydrated material to coat added dry gum with an impervious film which results in "fisheyes" with dry powder core centers.

2.   Dry blend the gum with a material such as dry sucrose or dextrose prior to addition into agitated water. Sugar physically separates the gum particles and encourages homogeneous dispersion of the gum and subsequent smooth solutions.

3.   Slurry the guar gum in an organic solvent such as glycerine, propylene glycol, or ethanol. The slurry is added to water with agitation.
      The hydration of guar gum in saturated sugar solutions (67 or 71%) is insignificant.[13] A saturated sucrose solution will disperse 4 to 5% guar gum and remain a fluid liquid. Sugar dispersion of guar gum is a convenient method for handling guar in highly automated and batch systems. As a sugar-guar dispersion is diluted with water, the dispersed guar hydrates smoothly.

4.   Use a disperser funnel, water circulated at a high shear rate, and a mixing tank. Pressure differential induced through rapid water flow promotes dispersion of gum into fast moving water. Finer grind gums require dry sugar dilution for this operation. Flow of dry powders into water must be controlled via mixing valve to maintain an optimum gum-water ratio. The mixing tank must be constantly agitated with a mechanical mixer until the dispersed gum particles are fully hydrated.

B. Viscosity

The most important single feature of guar gum is its water control or water-binding capacity usually expressed as viscosity. Viscosity measurements may be determined with a Brookfield or Haake viscometer.

In dilute solutions up to 0.4%, viscosity increases arithmetically; thereafter incremental guar increases have an exponential effect (Figure 3).

A 1% solution of medium fine grind guar, on an "as is" basis, has a viscosity of 3800 cps, at 25°C. The same guar on a moisture-free basis has a viscosity of 5500 cps.[14] Under typical conditions, doubling the concentration from 1 to 2% gives a tenfold increase in viscosity.

Guar gum solutions prepared for accurate viscosity measurements should be prepared on a moisture-free basis to eliminate the effect of possible moisture variables.

1. Particle Size

Fine grind guar gums tend to give higher final viscosities as compared to medium and coarse grinds.[16]

Field or plant use procedures frequently dictate a particular particle size selection. Coarse grind guars have a slower hydration rate than fine grind counterparts.

Field or plant circumstances seldom duplicate the conditions or circumstances of laboratory compounding. Rapidly hydrating hydrophilic colloids with improper dispersion will participate in the formation of clumps of undissolved gum. The clump surface will hydrate, forming a film or layer which prevents wetting of the interior. Once the clumps are formed, subsequent dispersion is extremely difficult.

2. Rheology

Guar gum solutions of 1% concentration or higher are thixotropic with thixotropy decreasing below the 1% level. Solutions are slightly thixotropic at 0.3% concentration, showing slight decreases in viscosity with increased shear.

3. Temperature

Like most gums, the viscosity of a fully hydrated guar solution will vary inversely with

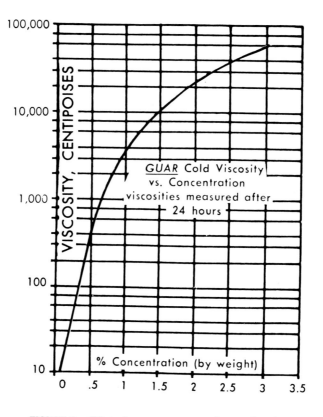

FIGURE 3.   Effect of guar gum concentration on viscosity.

temperature. A viscosity of 4000 cps at 20°C will become 2200 cps upon warming to 65.5°C. Guar gum solutions prepared at 25 to 40°C develop maximum viscosity.[10] Aqueous solutions of guar held for 1 hr at 85°C maintain constant viscosity.[15]

For guar systems preheated at 85°C followed by heating at 121°C for 15 min, adding 5 or 10% sugar provided protection against viscosity decrease.[13]

4. Effect of pH

Maximum hydration rate of guar is at pH 8.0 and the minimum rate is at pH 3.5 (Figure 4). The pH does not affect the final viscosity, even though the hydration rate does vary widely. Guar solutions are usually stable at pH 3.5 to 9.0 for extended periods. Other variables such as temperature, ionic strength, and other solutes may affect stability over an extended storage.

In cooking tests, Carlson et al.[15] combined 1 g guar with 99 mℓ of water, heated to 85°C and then cooled to 25°C to achieve a viscosity of 6700 cps. The same experiment with vinegar substituted for water resulted in a final viscosity of 1650 cps.

5. Effect of Electrolytes

Since the guar molecule is nonionic, excellent electrolyte compatibility is achieved over a broad ionic strength range.[17] Under controlled circumstances, borax, chromium, zirconium, calcium, and aluminum will insolubilize guar; however, these pH conditions and salt concentration are not applicable or allowed in foods.

Sequestering agents capable of inactivating iron ions, when added to guar solutions, improve stability. The additives of choice are citric, tartaric, and orthophosphoric acids in an amount of 0.25 to 0.5%, based on the weight of guar.[18]

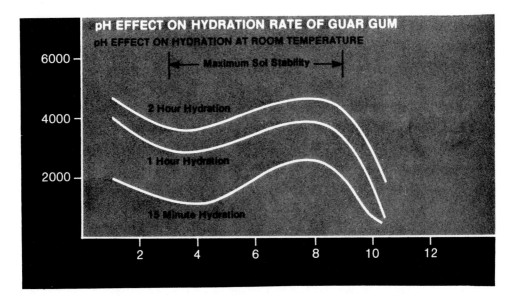

FIGURE 4.   Effect of pH on guar hydration rate at room temperature.

6. Biodegradation

Guar solutions, in common with most hydrophilic colloids, are subject to biodegradation. At 25°C, the viscosity of an unpreserved solution declines markedly on aging for 1 week. Food grade preservatives such as sodium benzoate, methyl and propyl parabens are effective in insuring molecular integrity.

7. Synergy

In cooking and autoclaving tests, fresh egg yolks, wheat starch, and fluid skim milk each combined separately with guar showed that the paired materials viscosity was higher than the single materials as tested.[15]

8. High-Viscosity Guar

Food grade guars (on an "as is" moisture basis) yielding a 1% solution viscosity in excess of 4000 cps are commercially available. By means of special processing the proportion of galactomannan fraction is increased and that of protein decreased.[13]

VII. APPLICATIONS

In comparison to other commercially available hydrocolloids (other natural gums, cellulose derivatives, algin derivatives, or xanthan gum), guar gum at equivalent solids in water has the highest viscosity.[10] Comparative viscosities of a 1% guar sol and 1.2% starch sol were 3200 cps and 100 cps, respectively.[20]

The designation hydrocolloid is a superlative term for describing guar gum. Guar gum modifies the behavior of water in food systems in a highly efficient manner. It reduces and minimizes friction in food products, thereby aiding processing and palatability of foods. Guar viscosity aids in the control of crystal size in saturated sugar solutions.

Perusal of food ingredient labels reveals many of the actual applications. Some of the most common uses are listed in Table 2.

Table 2
GUAR FOOD APPLICATIONS

Products	Examples
Dairy	Ice cream, ice milk, sherbet, ices, low fat soft serve, milk shakes, cottage cheese dressing, processed cheese, cheese dips
Pet foods	Dry and canned pet foods
Baked goods	Cakes, cake icing, cheese cake, pizza
Packaged dry mixes	Cake mix, salad dressing mix, breadings, instant soups, instant snacks, instant cereal
Condiments	Barbecue sauce, cocktail mix, relishes, taco sauce
Beverages	Juices, nectar, syrups

A. Dairy

1. Frozen Desserts

Ice cream products and the like are pasteurized by vat methods or high-temperature, short-time processes, 160°F for 30 min, or 175 to 180°F for 20 to 30 sec, respectively. Under traditional pasteurization conditions, guar gum is fully hydrated.

Guar imparts smoothness to ice cream by promoting small ice crystals during the freezing process. Small ice crystal size is maintained despite temperature fluctuations encountered in transport and in the automatic defrost home refrigerator. Guar also contributes body or bite resistance to ice cream thereby improving eating quality.

An ice cream mix is about 62% water, the solids portion consisting of fat, soluble sugars, proteins, hydrophilic colloids, and emulsifiers. The water of the ice cream mix becomes bound water by the mechanism of hydrogen bonds between the hydroxyl groups of colloid and water. Thus, viscosity is greatly increased, which slows the diffusion of solutes throughout. As ice cream mix becomes frozen, the bound water enhances and increases body.

Modern ice creams are frequently supersaturated with lactose due to the use of whey solids, yet the defect of crystallized lactose, also known as sandiness, is seldom encountered. The extremely high viscosity of the frozen matrix precludes the formation of detectable, undesirable, large lactose crystals.[21] Lactose or ice crystals greater than 20 μm are detectable by the consumer and perceived as a form of coarseness.[22]

Guar-based proprietary stabilizers encourage desirable low viscosity chocolate mixes.[23] For a typical 10% butterfat ice cream, the guar gum concentration is 0.12%. For incremental increases of 2% butterfat over 10%, gum is decreased by 0.02%; for incremental decreases of 2% butterfat, guar is increased by 0.02%. Thus a 4% butterfat ice milk contains 0.18% guar, and a 16% butterfat ice cream contains 0.06% guar.

Guar gum is usually compounded with other colloids and emulsifiers for sale to the dairy industry as a proprietary product.

Guar also controls ice crystal size and ensures smooth texture in ice milk, sherbet, Italian ices, and quiescently frozen ice pops.

2. Cottage Cheese

Cottage cheese is consumed as a mixture of about 62% curd and 38% dressing. Modern cottage cheese operations use mechanical means to move the finished product from the manufacturing point to packaging. In this kind of operation, dry curd is shattered and becomes unsightly. Guar gum in the dressing promotes curd integrity by friction reduction or lubricity, which allows the curds to slip during processing. Curd that is lubricated with dressing is easier to eat than dry curd. Guar also controls free water in the dressing yielding a homogeneous finished product with good storage characteristics.

The acceptable organoleptic response of guar solution is directly related to certain rheological properties. In an organoleptic evaluation of several gums and starches, guar was rated "very slight" to "somewhat slimy" at 0.6% concentration by Szczesniak et al.[24] (See Table 6, Chapter 1.) Guar, tragacanth, carrageenan, and karaya were categorized as "slightly slimy"; and pectin, methyl cellulose, carboxymethylcellulose, sodium alginate, and locust bean gum were called "slimy".[2,24]

3. Processed Cheese Products

Guar gum has found numerous applications in various soft cheese products. Cold pack cheese benefits from elimated syneresis or weeping, and more uniform textures and flavors are realized due to the control of moisture and migration.[25] Free moisture is controlled and texture enhanced in pasteurized processed cheese spreads and dips by the addition of guar. For soft cheeses, the use of guar increases yield, aids drainage, and imparts a soft, compact, tender texture to recovered curd.

A low-calorie imitation cheddar cheese loaf composition utilizing guar as part of the hydrocolloid system was prepared by Richardson.[26]

4. Pet Foods and Meat Products

Canned meat products are functionally improved by the use of about 0.5% guar based on total batch weight. Processing advantages are experienced such as: (1) less bumping and splattering during the cooking process; (2) easier pumping of the finished product from the vat to fillers; (3) more accurate filled weight control from reduced splashing; (4) smoother fill which keeps cans cleaner for labeling; (5) homogeneous dispersion of solids in liquid phase throughout filling process.

In the finished goods, the advantages are as follows: (1) fat is prevented from migrating during storage; (2) water and solids are maintained in a homogeneous state, thus eliminating syneresis during storage; (3) a reduced tendency for product voids in the can.[27]

Guar gum also imparts a desirable gloss or sheen to canned pet foods. Removal of the pet food from the can is also facilitated by the friction reduction function of guar.

5. Baked Goods

Guar influences baked goods in both the dough and finished product. Glicksman[28] pointed out that bread staling is at least partially related to moisture content and moisture transfer. The maximum use level for guar in baked goods is 0.35%.

A well-mixed dough containing guar has excellent filming properties; consequently, low gluten flours are benefited by guar addition. Rolls and bread augmented with guar have a soft texture and improved shelf life. In addition, the hinge of rolls is improved. The addition of guar gum to experimental gluten bread loaves show an increased volume upon baking.[29] In cake and biscuit dough, guar imparts softness and moister products that are more easily released from molds and less prone to crumbling while slicing. For each pound of guar added to a bread or roll batch, 6 lb of water should be added for hydration.

6. Packaged Dry Mixes

The versatile functional applications of guar lends itself to a variety of dry mixes. In each application, specific physical properties are imparted to the end product, thickening and viscosity control are obvious added features; however, improved mouthfeel is often realized at levels below 0.1% based on the finished product.

Guar gum in dry cake mixes has been reported to offer several advantages: (1) it aids the one-step mixing procedure, (2) it gives shorter batter mix time, (3) it yields improved internal structure with less crumbling; (4) it aids in application of frostings and icings; (5) it prolongs shelf life due to improved moisture retention; and (6) it permits freezing of finished cake.

Temporary suspension of solids may be achieved by the careful use of guar. In rehydrated particulate cereals, guar maintains a coherent mass and aids wetting or hydration.

7. Beverages and Condiments

Fruit nectars, which are usually made of fruit concentrates or puree, sugar, water, ascorbic acid, and citric acid, frequently contain 0.1 to 0.2% guar based on total weight. In addition to improved mouthfeel, the added guar facilitates the homogeneous dispersion of pulp during the filling process and subsequent storage. The resistance of guar to breakdown under the low pH conditions inherent to fruit products is an important factor in this application. Other physical properties of guar of interest in beverages include the rather bland flavor of guar, rapid hydration rate, and viscosity under ideal cost performance conditions.

Burrell[30] found guar an effective replacement for tragacanth in pickles and relish sauces provided that rapid product cooling is achieved.

Guar in tomato-based sauces helps maintain a desirable color as well as imparting body and stabilizing the system. Barbecue sauces, spiced meat sauces, taco sauce, and pizza dressings frequently incorporate guar as a stabilizer and water control agent. Guar sauces flow freely from the bottle and have excellent adherence. The use of guar permits lower solids content with substantial cost savings in products not covered by Standards of Identity.

8. Miscellaneous

Green young guar beans are consumed as human food. A recipe for guar curry supplied by Bharati K. Ravel[31] follows:

Guar Curry (with or without dumplings)

1 lb of young, green guar beans
4 tablespoons of corn or peanut oil
1 teaspoon of sugar
1 teaspoon of lemon juice
$1/2$ teaspoon of paprika or chili powder
$1/2$ teaspoon curry powder
$1/3$ teaspoon salt
$1/4$ teaspoon Ajma seeds (thyme; Indian ajowan seeds; *Trachyspermum ammi*. Use leaves if seeds not available)
1 cup of water
$1/2$ teaspoon garlic or onion flakes may be added according to taste

Remove longitudinal fibers of guar beans by breaking the bean tip and appropriate pulling. Cut beans into length desired, 1 to $1^1/2$ in.

In a frypan or skillet, heat the cooking oil to a fragrance point. Add Ajma seeds followed by guar beans, stir. Add curry powder and other flavorings, mix well. Cover the pan and allow to simmer for 1 or 2 min, then add water and with lid closed simmer the curry for about 20 min or until tender. Potato or flour dumplings may be incorporated during this part of the curry, in which case more water and longer simmering may be required.

Other vegetables such as potatoes, pumpkin, or squash and condiments as tomato sauce, wine vinegar, or apple sauce can also be incorporated. Scrambled eggs, cheese pieces, or ground meat may benefit the curry.

This recipe serves four persons. The guar curry is served over a bed of hot rice and topped with hard boiled egg slices.

In the realm of human health, consumption of guar gum has been shown to have therapeutic value. A 36% reduction in serum cholesterol levels following ingestion of 36 g of guar daily for 2 weeks was observed by Jenkins et al.[32] Later, Jenkins et al.[33] fed 15 g of guar daily for 2 weeks to 10 adults with type 11 hyperlipidemia and reported a 10.6% reduction in mean serum cholesterol levels.

Blood glucose levels have also been controlled in normal and diabetic persons by feeding guar gum on a daily basis.[34]

Several years ago, a new product called Javatol was compounded with 87.5% instant coffee and 12.5% guar gum. Javatol was intended to satisfy hunger between meals for dieters on weight loss programs.[2]

Block[35] combined guar gum, partially degraded soy protein, gelatin, and sugar to create a novel chocolate chiffon dessert.

Guar gum is a useful aid as a foam stabilizer and enhancer. Foam-mat-dried coffee concentrate utilized guar as an effective foam stabilizer.[36]

Opie et al.[37] used guar to improve the volume and texture in a culinary mix for baked goods. Guar improves whipping properties, foam body, and stability of egg whites prepared for icings.[38] Ziegler[39] incorporated guar gum as an effective whip-time-reducing agent in egg white systems designed for baked products.

More recently Haber[40] created a dry mix composition containing guar and other hydrocolloids, which upon reconstitution with cold milk, produced a yogurt-like dessert.

REFERENCES

1. **Ravel, K. R.,** personal communications, 1979.
2. **Glicksman, M.,** *Gum Technology in the Food Industry,* Academic Press, New York, 1969, chap. 5.
3. **Matlock, R. S.,** *Guar Variety and Cultural Studies in Oklahoma,* 1950—1959, Oklahoma Agric. Expt. Sta. Processed Series P-366,1960, 1—37.
4. Celanese World, Guar: A Gum For All Reasons, Vol. 4, No.2., 1979, 24.
5. Stein, Hall & Co., Jaguar® — Guar Gum, Stein, Hall & Co., New York, 1962.
6. Food Chemicals Codex, 3rd ed. National Academy Press, Washington, D.C., 1981, 141.
7. National Technical Information Service, GRAS (Generally Recognized as Safe) Food Ingredients — Guar Gum, PB 221-216, Springfield, Va., 1972, 35.
8. *Fed. Regis.,* Vol. 39, No. 185, Sept. 23, 1974.
9. Title 21, Code of Federal Regulations, Part 100 to 199, 1978, 719.
10. **Goldstein, A. M., Alter, E. N., and Seaman, J. K.,** *Guar Gum In Industrial Gums,* 2nd ed., Whistler, R. L., Ed., Academic Press, New York, 1973, 303.
11. Chemical Marketing Reporter, April 9, 1984.
12. **Hoyt, J. W.,** *J. Polymer Sci.,* Part B, 4, 713, 1966.
13. **Carson, W. A. and Ziegenfuss, E. M.,** The Effect Of Sugar On Guar Gum As A Thickening Agent, *Food Technol.,* 19, 64, 1965.
14. Henkel, Food Ingredients Division, Super Col® guar locust bean gums bulletin, 11 pages.
15. **Carlson, W. A., Ziegenfuss, E. M., and Overton, J. D.,** Compatibility And Manipulation Of Guar Gum, *Food Technol.,* 16, 50, 1962.
16. Hercules Inc., Guar Bulletins 957 and 958, Wilmington, Del.
17. Stein, Hall & Co., Jaguar® Guar Gum & Guar Derivatives, 31 pages.
18. **Jordan, W. A.,** U.S. Patent 3,007,879, 1961.
19. Kelco, Division of Merck & Co., Inc., Xanthan Gum, A Natural Biopolysaccharide For Scientic Water Control, 2nd ed., 1976, 36.
20. **Meer, G., Meer, W. A., and Tinker, J.,** Water-soluble gums, their past, present, and future, *Food Technol.,* 29(11), 22—30, 1975.
21. **Nickerson, T. A.,** Lactose crystallization in ice cream. IV. Factors responsible for reduced incidence of sandiness, *J. Dairy Sci.,* 45, 354, 1962.
22. **Keeney, P. G.,** Confusion over heat shock, *Food Eng.,* 116, June, 1979.
23. **Sperry, G. D.,** Algin stabilizers in chocolate mixes, *Ice Cream Rev.,* 37(3), 74, 1953.
24. **Szezesniak, A. S. and Farkas, E. H.,** Objective characterization of the mouthfeel of gum solutions, *J. Food Sci.,* 27, 381, 1962.

25. **Klis, J. B.,** Woody's Chunk O'Gold cold-pack cheesefood weeps no more, *Food Process. Mark.,* 27(11), 58, 1966.

26. **Richardson, T. W.,** U.S. Patent 4,089,981, 1978.

27. General Mills Bulletin GG-46, Super Col® In Canned Meat Products, 1964.

28. **Glicksman, M.,** The importance of hydrophilic gum constituents in processed foods, *Food Technol.,* 19(6), 53, 1965.

29. **Cawley, R. W.,** The role of wheat flour pentosans in baking. 11. Effect of added flour pentosans and other gums on gluten-starch loaves, *J. Sci. Food Agric.,* 15, 834, 1964.

30. **Burrell, J. R.,** Pickles and sauces, *Food Manuf.,* 33, 10, 1958.

31. **Ravel, B. K.,** personal communication, 1979.

32. **Jenkins, D. J. A., Newton, C., Leeds, A. R., and Cummings, J. H.,** Effect of pectin, guar gum, and wheat fibre on serum-cholesterol, *Lancet,* 1, 1116, 1975.

33. **Jenkins, D. J. A., Leeds, A. R., Slovin, B., Mann, J., and Jepson, E. M.,** Dietary fiber and blood lipids: reduction of serum cholesterol in type II hyperlipidemia by guar gum, *Am. J. Clin., Nutr.,* 32, 16, 1979.

34. **Jenkins, D. J. A., Leeds, A. R., Gasull, M. A., Cochet, B., and Alberti, K. G. M. M.,** Decrease in postprandial insulin and glucose concentrations by guar and pectin, *Ann. Intern. Med.,* 86, 20, 1977.

35. **Block, H. W.,** U.S. Patent 2,983,617, 1961.

36. **Hart, M. R., Graham, R. P., Ginnette, L. F., and Morgan, A. I., Jr.,** Foam for foam-mat drying, *Food Technol.,* 17, 1302, 1963.

37. **Opie, J. W. and Johnson, E. W.,** U.S. Patent 3,161,524, 1964.

38. **Sebring, M.,** U.S. Patent 3,378,376, 1968.

39. **Ziegler, H. F.,** U.S. Patent 3,711,299, 1973.

40. **Hober, G. J.,** U.S. Patent 4,081,566, 1978.

Chapter 7

TARA GUM

Martin Glicksman

TABLE OF CONTENTS

I. BACKGROUND

Although guar and locust bean gums are the two major galactomannans used widely in the food industry, a third member of the family, tara gum, with properties intermediate between guar and locust bean gums, has frequently been considered for potential use in the field. Tara gum is a related seed gum available at a comparatively low cost and offers an economic incentive for industrial exploitation.

Tara gum is derived from the endosperm of the tara seed of the leguminous plant *Cesalpinia spinosum* which is found in Peru and neighboring areas.[1] Structurally it is a galactomannan polymer consisting of a main chain of (1→4) linked β-D-mannopyranose units with side chains of α-D-galactopyranose linked (1→6) approximately every third unit. This ratio of mannose to galactose is intermediate between that of locust bean gum (1:4) and that of guar (1:2).[1,2]

II. MANUFACTURE

Much like the manufacture of guar and locust bean gums, tara gum is also produced by the mechanical crushing of the seed from the *Cesalpinia spinosum* pod and the physical separation of the gum from the endosperm portion. Actually, the whole tara pods were used for the production of tannin for tanning purposes with the tara gum being a by-product of the operation.[1,3]

III. SPECIFICATIONS

Although not permitted in food applications, refined tara gum has been sold commercially with the following specifications:[4]

	Percentage
Gum content	82.24
Protein	3.13
Acid-insoluble residue	1.28
Ash	0.98
Fat	0.37
Moisture	12.00

Viscosity of 1% cooked solution is 4100 cps.

IV. REGULATORY STATUS

Tara gum has not been approved for use in foods in the U.S., but it had been used in some cosmetic lotions from 1973 to 1978.[2]

Because of its functional similarity to locust bean gum and guar gum, it has been periodically considered as a less expensive substitute for these gums. Since tara gum had never been tested for potential carcinogenicity, it was included in the bioassay program of the National Toxicology Program of the U.S. Dept. of Health and Human Services.[2] A carcinogenic bioassay was conducted by feeding diets containing 25,000 or 50,000 ppm of tara gum to 50 F344 rats and 50 B6C3F1 mice of either sex for 103 weeks (2 years). Under conditions of this bioassay, tara gum was found to be noncarcinogenic.[2]

V. STRUCTURE AND COMPOSITION

Analysis and comparison of tara gum with other leguminous seeds by Anderson[1] showed

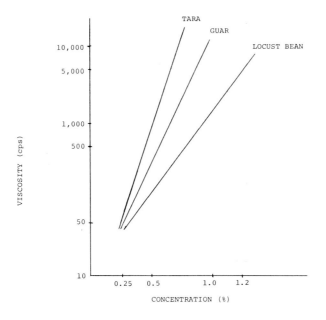

FIGURE 1.   Comparative viscosities of galactomannan gums. (From Wada,
K., Wada, K., and Deguchi, K., Japanese Patent 48-35463, 1973.)

that the endosperm portion of the tara seed was about 27% of the total weight of the seed;
this in turn yielded a soluble mucilage of tara gum (24%) and about 3% insoluble endosperm.
Further analysis of the tara gum mucilage yielded 70.9% anhydromannose, 26.3% anhy-
drogalactose, 0.5% acid insoluble material, 0.2% ash, and 0.0% pentosan.

Tara gum is more closely related to locust bean gum than guar gum, in that tara gum
behavior is so similar to that of fractions of locust bean gum of high galactose content that
a block structure seems likely. In locust bean gum, the galactose substituents are clustered
mainly in blocks of about 25 residues (hairy regions) interspersed by longer regions of
essentially unsubstituted mannan backbone (smooth regions) with the overall ratio of mannose
to galactose being about 4:1. Tara gum is believed to be similar in structure but with an
approximate 3:1 ratio of mannose to galactose.[5]

VI. PROPERTIES

Tara gum is fairly soluble in cold water, but for complete solution, heat must be applied,
much like locust bean gum. Aqueous solutions are neutral and highly viscous, and it has
been reported to be significantly higher in viscosity than guar or locust bean gum. Comparative viscosities, reported by Wada et al.,[3] are shown in Figure 1.

Weak gels can be formed by the addition of 60 to 70% sugar, and like locust bean gum,
it can also form a gel by interaction with xanthan gum, but these are weaker than xanthan-
locust bean gum gels.[5] The melting points and setting points of the tara-xanthan gels are
also slightly lower than those of locust bean-xanthan gels.[5] Tara gum also has a synergistic
gel-strength enhancing effect when combined with carageenan and agar gels.[3]

In general, tara gum has high viscosity, high water-holding ability, effective protective
colloidal characteristics and shows interfacial tension activity. In addition, it also has fairly
high acid resistance and salt resistance.[3]

VII. FOOD APPLICATIONS

Because of its structural similarity to locust bean and guar gums, tara gum was considered to be a potentially less expensive substitute for these gums in a number of thickening and stabilizing food applications. Some of the more interesting food applications are the following:

A. Dairy Products

Ice cream — Tara gum imparts good overrun characteristics (unlike locust bean gum) and gives ice cream a smooth, dense texture, desirable shape-retention properties, and good heat shock resistance.[3] Wada et al.[3] used a tara gum-carrageenan (9:1) stabilizing system at 0.3% to give a good quality ice cream. For sherbet, a tamarind seed gum-tara seed gum (4:1) blend at 0.3% was used. Dea and Finney[6] proposed the use of tara gum as part of a complex ice cream stabilizer system that also included locust bean gum, xanthan gum, agar and kappa-carrageenan together with glycerol, to produce spoonable, stable ice cream at freezer temperatures.

Cheese — The lack of milk reactivity properties of tara gum makes it useful in the production of cheese where it can facilitate phase separation in milk systems.[3]

Milk puddings — Tara gum was used by Glicksman and Farkas[7] in the development of phosphate-free puddings having improved textural stability properties. These puddings were stable, with no syneresis, over wide temperature ranges from freeze-thaw processing to high temperature canning procedures.

B. Bakery Products

In sponge-cake, baked cake, or bread, tara gum imparts softness, form retention, long-lasting freshness and easier slicing characteristics.[3]

C. Dessert Gel Products

A thermally-reversible, acid-stable, room-temperature setting water-dessert gel, that did not exhibit syneresis, was prepared by using tara gum in combination with xanthan gum.[8]

D. Sauces and Condiments

Because of its fairly high acid resistance and salt resistance, tara gum can be effectively used as a thickening and stabilizing agent in ketchup, soy sauce, soybean paste, and related condiment products.[3]

E. Miscellaneous

Processed seafoods — Tara gum in combination with xanthan gum is reported to markedly improve the emulsion stability and water retention.[9]

Frozen foods — Because of its easy dispersbility in cold water and its high viscosity and high water-holding ability, tara gum is an effective glazing agent.[3]

Pasta — Tara gum reportedly strengthens the structure of noodles and reduces breakage.[3]

REFERENCES

1. **Anderson, E.,** Endosperm mucilages of legumes — occurrence and composition, *Ind. Eng. Chem.,* 41(12), 2887—2890, 1949.
2. NTP, National Toxicology Program, NTP Technical Report on the Carcinogenesis Bioassay of Tara Gum, NTP TR 224, National Institute of Environmental Health Sciences, National Institutes of Health, Public Health Service, Dept. of Health and Human Services, Research Triangle Park, N.C., 1982.
3. **Wada, K., Wada, K., and Deguchi, K.,** Method for Improving the Quality of Food and Drink, Japanese Patent 48-35463, October 27, 1973.
4. **Duffy, J. J.,** private communication, Dycol Chemical Co., 1970.
5. **Dea, I. C. M., Morris, E. R., Rees, D. A., Welsh, E. J., Barnes, H. A., and Price, J.,** Associations of like and unlike polysaccharides: mechanism and specificity in galactomannans, interacting bacterial polysaccharides, and related systems, *Carbohydr. Res.,* 57, 249—272, 1977.
6. **Dea, I. C. M. and Finney, D. J.,** Stabilized Spoonable Ice Cream, U.S. Patent 4,145,454, March 20, 1979.
7. **Glicksman, M. and Farkas, E. H.,** Pudding Compositions, U.S. Patent 3,721,571, March 20, 1973.
8. **Glicksman, M. and Farkas, E. H.,** Gum Gelling System Xanthan-Tara Dessert Gel, U.S. Patent 3,784,712, Jan. 8, 1974.
9. San-Ei Chemical Industries, Ltd., Emulsifiers for processed marine foods, Jpn. Tokyo Koho JP 58-00,309 (83-00,309), 1983.

Chapter 8

TAMARIND SEED GUM

Martin Glicksman

TABLE OF CONTENTS

I. BACKGROUND

Tamarind gum, also known as tamarind kernel powder (TKP), tamarind seed powder, and tamarind endosperm powder, is obtained from the endosperm of the seeds of the tamarind tree, *Tamarindus indica.* The tamarind tree, a member of the evergreen family, is one of the most important and common trees of southeast Asia and widely indigenous to India, Bangladesh, Burma, Sri Lanka, and Malaysia; it is also found in many other parts of the world such as Formosa, Egypt, and Florida.[5,6]

Like many other developments, tamarind gum was discovered during a search for new sizing materials during shortages caused by World War II. It came into commercial production in 1943 as a replacement for starch in cotton sizing used in Indian textile mills.[1] In the U.S., its major industrial use has been as a wet end additive in the paper industry as a replacement for starches and galactomannans. It is often used more effectively in combination with other gums such as guar and alginate.[23]

Tamarind gum is not approved for food use in the U.S., but has been used in foods in India, where most of it is produced. Currently, purified, refined tamarind gum is produced and permitted in Japan as a thickening, stabilizing, and gelling agent in the food industry. The market for this material in Japan is reported to be in the vicinity of 800 metric tons annually.

The tamarind tree, *Tamarindus indica,* is a large, rapidly-growing evergreen tree that lives for more than 100 years. A full-grown tree is about 80 ft tall with a circumference of about 25 ft. It starts to bear fruit after about 13 to 14 years and continues to yield abundant crops (about 400 to 500 lb of fruit) for more than 60 years.[5,6]

The fruit is a large flat pod, about 4 to 6 in long and contains seed, stringy fibrous matter, and acidic pulp. A rough analysis by Hooper[5,7] reported the proportions to be 33.9% seed, 55.0% pulp, and 11.1% shell and fiber.

The pulp portion of the fruit is used as the chief acidifying agent in Indian curries, chutneys (chatnis), and sauces, while the tamarind seed is a by-product of the tamarind pulp industry and the source of the polysaccharide gum or hydrocolloid. About 150,000 tons of seed are produced annually.[25]

The tamarind seed has a flat, irregular shape, being round, oval, or four-sided. The length of a side is about 0.6 in. and the thickness is about 0.3 in. The seed has an outer brown seed coating or hull or testa comprising about 30% of the weight and covering a whitish to tan colored, cereal-like kernel which accounts for 70% by weight.[5] The seed coating or testa can be removed by roasting the seed to make the testa brittle or by soaking in water to loosen the seed coat.

The isolated tamarind seed kernels resemble cereals and have the following approximate composition (d.b.):[5]

	Percentage
Protein	15.4—22.7
Oil	3.9—7.4
Crude fiber	0.7—8.2
Nonfiber carbohydrates	65.1—72.2
Ash	2.45—3.3

The functional tamarind gum or hydrocolloid is a polysaccharide polymer that comprises a major portion of the dehulled tamarind kernel. It is commercially available as the crushed, powdered kernel powder sold as tamarind seed gum, TKP, or tamarind flour. This powdered tamarind seed powder usually contains at least 50% of the tamarind polysaccharide, the active hydrocolloid principal.

The pure tamarind hydrocolloid is commercially available as a water extract of the TKP that is purified, dried, and ground to a fine powder and sold under the trade name Glyloid by the sole manufacturer of this material, Dainippon Pharmaceutical Co., Ltd., in Japan.[2]

In all experimental work or references pertaining to the properties and applications of tamarind seed hydrocolloid, it is critically important to be specific about the material used, i.e., (1) whether it is crude tamarind kernel powder containing about 50% gum or (2) purified, tamarind seed extract comprising 100% tamarind gum polysaccharide.

In this chapter, the crude (50%) material will be referred to as tamarind kernel powder (TKP), tamarind seed gum, flour, or powder, while the purified (100%) material will be called tamarind seed extract or tamarind seed polysaccharide.

II. MANUFACTURE

Tamarind seed flour and gum manufacture is similar to that for guar, locust bean gum, and other seed gums, entailing primarily the separation and grinding to a powder of the endosperm portion of the seed.

Basically the process begins by thoroughly washing the seeds with water to free them from the attached pulp. The seeds are then heated to above 150°C for at least 15 min to make the testae or seed coatings brittle and friable. The testae are then removed and the seeds decorticated to leave the heavier, crushed endosperm, which is then ground to yield commercial tamarind seed powder or tamarind seed gum.[1]

This material which contains much insoluble matter, can be further purified to give the pure tamarind polysaccharide gum. The process is essentially a water extraction whereby the tamarind seed powder is boiled and agitated with about 30 to 40 times its weight of water for about 30 to 40 min and then allowed to sit overnight in a settling tank to allow the protein and fiber to precipitate and settle out. The supernatant liquor is concentrated to about half its volume, mixed with filter-aid, and filtered through a filter-press. The purified liquid is then drum-dried and ground to yield a purified tamarind seed extract.[1] Additional purification procedures involving precipitation with salts and washing with alcohol have also been developed.[14,15] Purified tamarind seed extracts of this type are currently being manufactured by the Dainippon Pharmaceutical Co., Ltd., in Japan and sold under the trade name of Glyloid.[2]

III. SPECIFICATIONS

A. Tamarind Kernel Powder

Unofficial specifications for crude TKP or tamarind seed gum have been outlined by Rao and Srivastava[5] as follows:

- Creamy white in color with no disagreeable odor
- Free of testae
- Mesh size of 100% through 85 mesh and at least 80% through 100 mesh (British Standard Test Mesh)
- Relative viscosity of 4.5 to 5.0 cps (0.5% solution at 35°)
- Not more than 3.0% ash
- Not less than 50% polysaccharide
- D-Xylose content of 19 to 21%

Meer Corp.[12] has also reported a viscosity specification of a 2% solution to be 170 to 250 cps.

Table 1
PRODUCT SPECIFICATIONS FOR TAMARIND SEED EXTRACTS[2]

Item	Quality standard	
	Glyloid 3A	Glyloid 3S
Viscosity, 1.5% soln	550—800 cps	500—800 cps
Gel strength, 1.0% concentration	Not less than 150 g/cm²	Not less than 150 g/cm²
Arsenic	Not more than 1 ppm	Not more than 1 ppm
Heavy metals	Not more than 10 ppm	Not more than 10 ppm
Protein	Not more than 3%	Not more than 3%
Fat	Not more than 1%	Not more than 1%
Loss on drying	Not more than 7%	Not more than 7%
Residue on ignition	Not more than 5%	Not more than 5%
Bacteria count	Not more than 2,000/g	Not more than 10,000/g
Yeast and mold count	Not more than 300/g	Not more than 1,000/g
Coliforms	Negative	Negative
Heat-resistant bacteria	Not more than 1,000/g	Not more than 1,000/g
Pathogenic bacteria	Negative	Negative

B. Tamarind Seed Polysaccharide

The purified water extract of TKP is a complex polysaccharide consisting of D-galactose, D-xylose and D-glucose in a molar ratio of 1:2:3.[24] It is available commercially in two grades — Glyloid 3S, which forms viscous solutions in cold water without heating, and Glyloid 3A, which is insoluble in cold water and requires heating at 75°C for at least 15 min for solubilization.

Product specifications for these two tamarind gum extracts are shown in Table 1.[2]

IV. REGULATORY STATUS

Tamarind seed flour or gum is not permitted in food in the U.S. at the present time, but the purified tamarind seed extract (Glyloid) has been used in Japan since 1964 as a thickener, gelling agent and stabilizer in various kinds of processed foods.[2,3]

Dainippon Pharmaceutical Co. conducted 2-year feeding toxicity tests on tamarind seed polysaccharides and reported the following results:[2,4]

"Tamarind seed polysaccharide (Glyloid) was incorporated at the level of 4, 8, 12% in a standard commercial diet and fed ad lib, to male and female rats for 2 years. No significant changes were noted in the behavior, mortality, body weight, food intake, biochemical analysis of urine and blood, hematological test, organ weight and histopathological findings of rats receiving Glyloid. In all groups including control group, spontaneous diseases with aging, such as myocardial change, nephropathy, mammary tumor (in female), pituitary tumor, etc. were seen. Deaths of the animals in this study were mainly attributable to these spontaneous diseases."[4]

V. STRUCTURE

Tamarind seed gum is a polysaccharide polymer made out of D-galactose, D-xylose, and D-glucose and has a molecular weight of about 115,000. The proposed structure shown in Figure 1 projects a backbone of glucose with branching on three out of every four glucose units, of a xylose side chain linked also to a galactose molecule.[2] Some studies have also indicated the presence of small amounts of L-arabinose,[13] but basically the exact structure has still to be defined.

VI. PROPERTIES

A. Viscosity

TKP disperses and hydrates quickly in cold water, but does not reach maximum viscosity

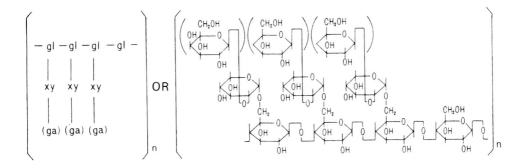

FIGURE 1. Structure proposed for tamarind seed polysaccharide. (Courtesy of Dainippon Pharmaceutical Co., Inc., Osaka, Japan.)

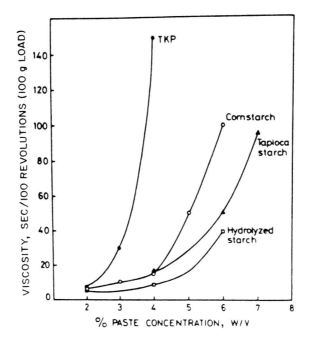

FIGURE 2. Viscosity as a function of concentration of solutions of TKP compared to starches. (From Rao, P. S. and Srivastava, H. C., *Industrial Gums*, 2nd ed., Whistler, R. L., Ed., Academic Press, New York, 1973, 369.)

unless it is heated for 20 to 30 min.[5] The solution exhibits typical non-Newtonian flow properties common to most other hydrocolloids. In general, the viscosity of TKP dispersions is much higher than those of starch solutions of equal concentration, as shown in Figure 2.[5]

Tamarind seed extract or polysaccharide is more soluble but still requires heating to obtain maximum solution viscosity. A typical 1.5% gum solution will yield a viscosity of 500 to 800 cps at 25°C.[2] Typical viscosity-concentration relationships are shown in Figure 3. The relationship of tamarind seed polysaccharide viscosity vs. shear rate is shown in Figure 4 and appears to indicate Newtonian flow behavior at low viscosity concentrations.

B. Effect of pH
Tamarind seed gum or polysaccharide has excellent stability over the acid pH range and

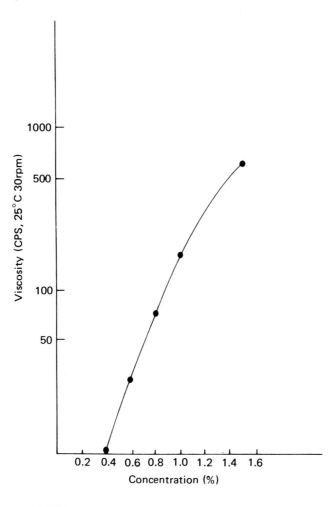

FIGURE 3.   Viscosity concentration relationship of tamarind seed extract. (Courtesy of Dainippon Pharmaceutical Co., Inc., Osaka, Japan.)

Maximum viscosity of TKP solutions was reported to be in the pH range of 7.0 to 7.5,[5] which is somewhat at variance with studies on the pure extract.

C. Effect of Electrolytes

Tamarind seed gum shows excellent stability in high concentrations of salt (20%) as shown in Figure 6.[2] Other electrolytes have some effect on viscosity of TKP solutions. Copper sulfate and zinc chloride reduce the viscosity while soluble calcium and magnesium salts have no appreciable effect; calcium chloride is even reported to cause an increase in viscosity.[5]

D. Effect of Temperature

Tamarind seed gum shows good stability to heat at lower temperatures (65°C) over a pH range 3.0 to 7.0. At higher temperatures (100 to 110°C), tamarind gum is stable at neutral pH, but degrades rapidly at lower pHs, similar to many other gums.[2]

E. Effect of Sugars

Sugars have an unusual synergistic effect upon tamarind seed polysaccharide. As sucrose

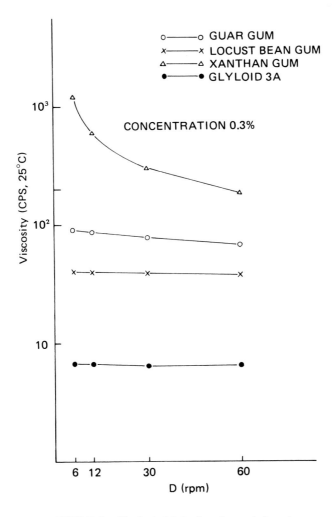

FIGURE 4. Rheological behavior of tamarind seed extract.
(Courtesy of Dainippon Pharmaceutical Co., Inc., Osaka, Japan.)

or glucose is added to a solution of tamarind seed gum, the viscosity increases markedly (Figure 7), and when the concentration of sugar exceeds 40%, a gel forms.[2,3]

F. Gelation

Tamarind seed polysaccharide has the ability to form gels in the presence of sugar or alcohol. In this respect, it is similar to pectin and can be used to form pectin-like gels in such products as jams, jellies, marmalades, and preserves.[3] Good quality gels can be made using tamarind seed gum at concentrations of 0.5 to 1.5% with sugar concentrations ranging from 70% down to as low as 45%. If alcohol (up to 20%) is added, the amount of sugar needed to form a gel can be substantially reduced and even eliminated in some cases.[2,3]

To form a gel, heating is required in order to get all of the gum into solution and upon cooling to room temperature, the gel will form. Maximum gel strength is reported to be at pH 2.7 for 1.0% gum concentration.[3]

The relationship of gel strength to gum and sugar concentrations is shown in Figure 8[2] while the relationship of gel strength to sucrose and ethanol concentrations is shown in Figure 9.[2]

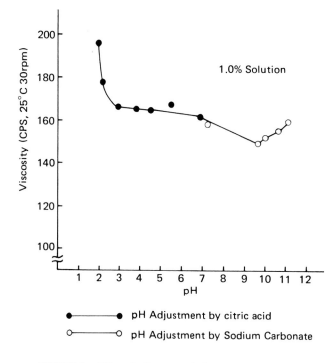

FIGURE 5.   Effect of pH on tamarind seed extract viscosities.
(Courtesy of Dainippon Pharmaceutical Co., Inc., Osaka, Japan.)

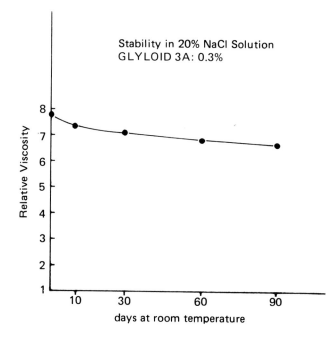

FIGURE 6.   Stability of tamarind seed extract in salt solution.
(Courtesy of Dainippon Pharmaceutical Co., Inc., Osaka, Japan.)

G. Film-Forming Properties

TKP films can be prepared by conventional pouring of dispersions onto a glass plate, spreading it into a thin layer, and allowing it to dry. The resultant films are smooth, strong,

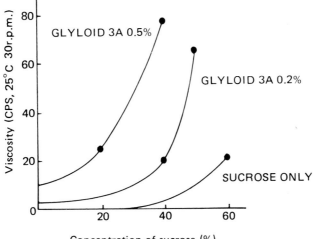

FIGURE 7. Synergistic effect of sucrose on tamarind seed extract viscosities. (Courtesy of Dainippon Pharmaceutical Co., Inc., Osaka, Japan.)

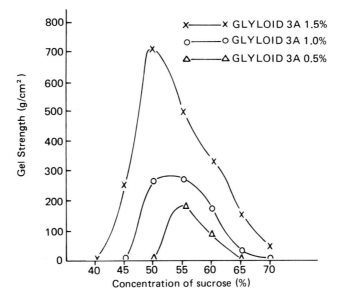

FIGURE 8. Relationship of gel strength to sucrose concentration. (Courtesy of Dainippon Pharmaceutical Co., Inc., Osaka, Japan.)

continuous, and extensible. They compare well with those of starch and appear to be more adhesive.[5]

VII. FOOD APPLICATIONS

In the U.S., tamarind seed gum or polysaccharide is not permitted in food products under current FDA regulations, probably because no user or supplier saw fit to petition for its use

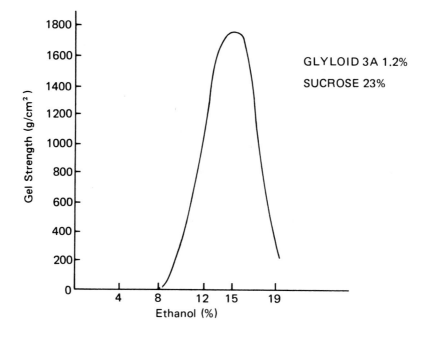

FIGURE 9.   Relationship of gel strength to sucrose and ethanol concentrations. (Courtesy of Dainippon Pharmaceutical Co., Inc., Osaka, Japan.)

Table 2
TAMARIND GUM FOOD APPLICATIONS IN JAPAN[2]

Food	Function	Use amount (%)
Filling (flour paste)	Syneresis inhibitor	0.1 ~ 0.3
Ice cream, sherbet	Stabilizer	0.1 ~ 0.2
Sauce, gravy	Thickening agent	0.2 ~ 0.3
Tomato ketchup	Thickening agent	0.1 ~ 0.2
Canned food	Thickening agent	0.1 ~ 0.5
Jam, marmalade	Gelling agent	0.1 ~ 0.5
Mayonnaise, salad dressing	Stabilizer	0.1 ~ 0.3
Fruit juice	Thickening agent	0.1 ~ 0.3
Cake mixes	Thickening agent	0.1 ~ 0.3
Powdered soup	Thickening agent	0.1 ~ 0.5
Noodle	Improving texture	0.1 ~ 0.5
Pickles	Thickening agent, syneresis inhibitor	0.1 ~ 0.5

as a food additive. In Japan, however, tamarind seed gum has been used for many years as a food additive or ingredient and is now used widely in many food applications as listed in Table 2.[2]

A. Jams and Jellies

One of the more important characteristics of tamarind seed gum is its ability to form jellies with concentrated sugar solutions over a wide pH range. Thus it has been used as a substitute for fruit pectin in jam, jelly, and preserve products.[5] Early work comparing tamarind seed gum and pectin in such applications reported advantages of tamarind gum. Much less tamarind gum (about half) is required to form a fruit gel of equal gel strength to a pectin gel. In addition, tamarind gum can form gels over a wide pH range (approximately 2 to 10), whereas pectin can only gel within a comparatively short acid pH range. Another advantage is that tamarind seed gum is not affected by boiling in neutral aqueous solutions, whereas pectins undergo severe degradation on boiling under such conditions. Both tamarind seed gum and pectins are readily degraded by treatment with hot acids or hot alkalis.[5]

B. Confectionery

Tamarind seed polysaccharide or gum has been used in the preparation of jelly candies and related confections such as jujubes. Conventional methods are used wherein the hot jelly solutions are poured into starch molds and allowed to set. The molded pieces are then further dried to a desirable moisture content, rolled in sugar to form a surface coating of sugar, and then packaged.[5,8]

C. Salad Dressings and Mayonnaise

Tamarind seed gum is reported to be comparable to other gums such as tragacanth, arabic, and karaya in stabilizing oil emulsions (by reduction of interfacial tension).[17] It has thus been used as an emulsion stabilizer in mayonnaise and similar food emulsion products.[9-11]

The emulsion stabilizing properties of tamarind seed gum in oil-in-water emulsions can be improved by combining it with xanthan gum in approximately equal proportions.[18]

Comparative studies by Tanaka and Fukuda[22] compared tamarind seed extract favorably with pectin, tragacanth, and guar gums in the stabilization of French-style salad dressings.

D. Frozen Desserts

Tamarind seed polysaccharide alone or in combination with carrageenan, sodium alginate, or low-methoxy pectin is a very effective stabilizer in frozen dessert products such as ice cream, ice milk, sherbet, and water ice.[9,16] It has been reported to give good overrun, no wheying-off, excellent heat shock resistance, and good water-holding properties without the separation of ice crystals and sugar after long storage.[16]

Wada et al.[19] employed a blend of tamarind seed gum (4 parts) and tara gum (1 part) at 3% levels as a stabilizer for producing a smooth, shape-retaining, heat shock resistant sherbet.

E. Miscellaneous

Tamarind seed gum has been used to provide adhesiveness in protective coatings based on casein and starch hydrolyzates for preserving eggs and similar products.[20] It has been employed as a thickening agent in an improved kneaded dough, bread production system.[21]

In Japan, tamarind seed gum is also used as a thickening agent in condiments such as ketchup, soy sauce, and pickled seaweeds. It is also utilized in specialized pasta products such as Ramen noodles.

REFERENCES

1. **Gerard, T.,** Tamarind gum, in *Handbook of Water-Soluble Gums and Resins,* Davidson, R. L., Ed., McGraw-Hill, New York, 1980, chap. 23.
2. Dainippon Pharmaceutical Co., Inc., Glyloid (Tamarind Seed Polysaccharide) Bulletin, Dainippon Pharmaceutical Co., Ltd., Osaka, Japan, 1982.
3. Dainippon Pharmaceutical Co., Inc., How to Use Glyloid in Food Applications, Dainippon Pharmaceutical Co., Ltd., Osaka, Japan, 1982.
4. Dainippon Pharmaceutical Co., Inc., Tamarind seed polysaccharide feeding study, *J. Toxicol. Sci.,* 3, 163—192, 1978.
5. **Rao, P. S. and Srivastava, H. C.,** Tamarind, in *Industrial Gums,* 2nd ed., Whistler, R. L., Ed., Academic Press, New York, 1973, 369—411.
6. **Rao, P. S.,** Tamarind, in *Industrial Gums,* 1st ed., Whistler, R. L. and BeMiller, J., Eds., Academic Press, New York, 1959, 461—504.

7. **Hooper, D.,** *Agric. Ledger*, 2, 13, 1907 (see Reference 5).
8. **Rao, P. S.,** *Research India*, 4, 173, 1959.
9. **Shoji, O., Wada, K., Tamura, A., and Wada, K.,** Frozen Desserts, U.S. Patent 3,342,608, 1967; *Chem. Abstr.*, 68, 2121, 1968.
10. **Savur, G. R.,** *Indian Food Packer*, 9 (7), 15, 1955.
11. **Savur, G. R.,** Utilization of tamarind seed polyose in food industries, *Indian Food Packer*, 9 (2), 13, 1955.
12. Meer Corp., Tamarind Seed Gum brochure.
13. **Srivastava, H. C. and Singh, P. P.,** Structure of the polysaccharide from tamarind kernel, *Carbohydr. Res.*, 4, 326—342, 1967.
14. **Gordon, A. L.,** Tamarind Seed Polysaccharide Recovery, U.S. Patent 3,399,189, 1968; *Chem. Abstr.*, 69, 97856u, 1968.
15. Dainippon Pharmaceutical Co., Ltd., Jellose from Tamarind Seed, British Patent 1,007,303, 1965; *Chem. Abstr.*, 63, 17809e, 1965.
16. **Shoji, O., Wada, K., Tamura, A., and Wada, K.,** Frozen Desserts, U.S. Patent 3,342,608, 1967.
17. **Patel, R. P. and Raghunathan,** *Indian J. Pharm.*, 21, 159, 1959.
18. **Kyu-Pi Co.,** Stabilization of Food Emulsions, Japanese Patent 82—91,172, 1982; *Chem. Abstr.*, 71, 126011a, 1982.
19. **Wada, K., Wada, K., and Deguchi, K.,** Method for Improving the Quality of Food and Drink, Japanese Patent 48-35463, 1973.
20. Matsutani Kagaku Ko, Coating Compositions for Preserving Foods, Japanese Patent 51-56,580, 1981.
21. **Atsumi, S., Sasaki, M., and Kitamura, I.,** Method for Producing Bread, U.S. Patent 4,405,648, 1983.
22. **Tanaka, M. and Fukuda, H.,** Studies on the texture of salad dressings containing xanthan gum, *Can. Inst. Food Sci. Technol.*, 9(3), 130—134, 1976.
23. **Yin, R. I. and Lewis, J. G.,** Novel Blend of Algin, TKP and Guar Gum, U.S. Patent 4,257,816, 1981.
24. **Whistler, R. L.,** Polysaccharides, in *Encyclopedia of Polymer Science and Technology*, 11, 415, 1969.
25. **Cottrell, I. W. and Baird, J. K.,** Gums, in *Kirk-Othmer's Encyclopedia of Chemical Technology*, 3rd ed., Vol. 12, John Wiley & Sons, New York, 1980, 59—60.

Plant Extracts

Chapter 9

PECTINS

Steen Højgaard Christensen

TABLE OF CONTENTS

I. BACKGROUND

A. Introduction

Pectin is the designation for a group of valuable polysaccharides extracted from edible plant material and used extensively as gelling agents and stabilizers by the food industry. The world production of pectin in 1982 is estimated to amount close to 16,000 metric tons. Main application field is traditionally fruit-based products, especially jams and jellies, where pectin is used as a gelling agent. Unlike most other food hydrocolloids, pectin shows optimum heat stability at acidic conditions and is therefore a potential candidate whenever a texturizer or stabilizer is required in an acidic food product. Increasing amounts of pectin have accordingly found application outside the fruit processing industry. Some 1000 metric tons are today used in confectionery products and as stabilizers for acidic milk drinks, and new application possibilities are constantly being investigated and developed. A considerable amount of pectin is further used outside the food industry primarily for pharmaceutical purposes.

The word pectin stems from the Greek phrase πηχτος meaning ''to congeal or solidify''. The congealing properties of pectin have been utilized for centuries since the method of preserving fruit by boiling down with sugar to form a jelly was invented. The discovery of the chemical compound was made by Vauquelin[1] in 1790, but Braconnot[2] was the first to characterize it as the active fruit component responsible for gel formation and suggest the name ''pectin''. During the 19th century, after the initial characterization of pectin, a considerable amount of scientific research regarding the molecular structure and the biological and rheological properties of the hydrocolloid was made. The isolation of commercial pectins from suitable plant material commenced early in this century, and since then a vast amount of literature concerning the chemistry, production, and functional properties of pectins has been produced.

The book on pectin substances by Kertesz[3] published in 1951 provides a comprehensive review of the knowledge about pectin at that time and is still considered as the basic book on pectin. More recent reviews have been made by various authors.[4-12]

B. Origin

Like starch and cellulose, pectin is a structural carbohydrate product present in all plants. Pectic substances are integral components of the cell structures and play an important role as cementing material in the middle lamellae of primary cell walls. Pectic substances are abundant in fruits and vegetables and to a large extent responsible for firmness and form retention of their tissue. The process of ripening and maturing involves enzymatic hydrolysis and depolymerization of the parent pectic substances, partly to yield soluble pectins.

C. Nomenclature

Pectin and pectic substances are heteropolysaccharides mainly consisting of galacturonic acid and galacturonic acid methyl ester residues. In the early days of pectin research, a great deal of confusion regarding pectin terminology was created by the various investigators in the field. In 1944, the American Chemical Society adopted a ''Revised Nomenclature of the Pectic Substances'', which is still used by many scientists as a standard pectin terminology.[13] These uniform definitions are as follows:

- *Pectic substances* are those complex colloidal carbohydrate derivatives that occur in or are prepared from plants and contain a large proportion of anhydrogalacturonic acid units, which are thought to exist in a chain-like combination. The carboxyl groups of polygalacturonic acids may be partly esterified by methyl groups and partly or completely neutralized by one or more bases.
- *Protopectin* is the water-insoluble parent pectin substance that occurs in plants and which on restricted hydrolysis yields pectin or pectinic acids.
- *Pectinic acids* are the colloidal polygalacturonic acids containing more than a negligible proportion of methyl ester groups. Pectinic acids, under suitable conditions, are capable of forming gels in water with sugar and acid, or, if suitably low in methoxyl content, with certain ions. The salts of pectinic acids are either normal or acid pectinates.
- *Pectin* (or *pectins*) are those water-soluble pectinic acids of varying methyl ester content and degree of neutralization which are capable of forming gels with sugar and acid under suitable conditions.
- *Pectic acid* is a term applied to pectic substances composed mostly of colloidal polygalacturonic acids and essentially free from methyl ester groups. The salts of pectic acids are either normal or acid pectates.
- *Protopectinase* is the enzyme that converts protopectin into a soluble product. It has also been called ''pectosinase'' and ''propectinase''.
- *Pectinesterase* (PE), or *pectinmethylesterase*, is the enzyme that catalyzes the hydrolysis of the ester bonds of pectic substances to yield methanol and pectic acid. The name ''pectase'' does not indicate the nature of the enzyme action and has given way to these more specific names.
- *Polygalacturonase* (PG), or *pectin polygalacturonase*, is the enzyme that catalyzes the hydrolysis of glycosidic bonds between de-esterified galacturonide residues in pectic substances. ''Pectinase'' is frequently used to designate the glycosidase as well as pectic enzyme mixtures.

Due to increasing commercialization of pectinic acids with a low methyl ester content and partly amidated pectinic acids for use in various food products, a modified definition of the food additive, pectin, has been adopted recently by the food industry and food legislative authorities. The following definition complies with contemporary food legislation in most countries:

- *Pectin* is a complex, high molecular weight polysaccharide mainly consisting of the partial methyl esters of polygalacturonic acid and their sodium, potassium, and ammonium salts. In some types (amidated pectins) galacturonamide units further occur in the polysaccharide chain. The product is obtained by aqueous extraction of appropriate edible plant material, usually citrus fruits and apples.

II. DESCRIPTION

A. Classification of Commercial Pectins

Commercial pectin is generally obtained by dilute-acid extraction of citrus albedo or apple pomace followed by various purification and isolation processes. The product usually occurs as a practically odorless, off-white to yellowish white coarse to fine powder having a mucilaginous taste.

For practical purposes, the pectin molecule can be considered as an unbranched chain containing 200 to 1000 galacturonic acid units linked together by α-1,4-glucosidic bonds. Some of the galacturonic acid units in the molecule are esterified and present as galacturonic acid methyl esters. Remaining acid groups may be partly or fully neutralized to form ammonia, potassium, or sodium salts (Figure 1).

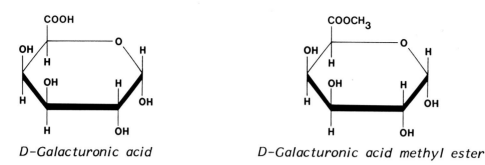

D-Galacturonic acid   D-Galacturonic acid methyl ester

FIGURE 1.   Principal units in the pectin molecule.

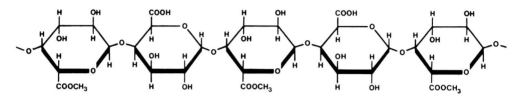

FIGURE 2.   Section of a high-ester pectin molecule with a degree of esterification ~60%.

The *degree of esterification* is defined as the ratio of esterified galacturonic acid units to total galacturonic acid units in the molecule. *High-ester pectins* — often called HM-pectins or high methoxyl pectins — are pectins with a degree of esterification above 50% (Figure 2). High-ester pectins require a minimum content of sugar (a soluble solid above approximately 55%) and acid (a pH around 3.0) in order to form gels. Once formed, the high-ester pectin gel cannot be remelted by heating. The degree of esterification of high-ester pectins determines their relative gelling rate. Pectins with a degree of esterification in the range 70 to 75% are often referred to as *rapid set* high ester pectins while the designation *slow set* is used for pectins with a degree of esterification in the range 55 to 65%. Ultrarapid set-, medium set-, and extra slow set- are prefixes sometimes used to characterize marginal or intermediate high-ester pectin types.

Low-ester pectins (LM-pectins or low methoxyl pectins) are pectins with a degree of esterification below 50%. Commercial low-ester pectins are generally produced from plant material containing high-ester pectin. The transformation (de-esterification) of high-ester pectin accordingly takes place at controlled conditions during the manufacturing process by treatment at either mildly acidic or alkaline conditions. If ammonia is used for the alkaline de-esterification process, a so-called *amidated low-ester pectin* will result. Apart from galacturonic acids and galacturonic acid methyl esters, amidated low-ester pectins contain galacturonamide units in the molecular chain (Figure 3).

The gelling mechanism of low-ester pectins differs essentially from that of high-ester pectins. To obtain gel formation in a system containing low-ester pectin, the presence of calcium ions is crucial. Low-ester pectins may, on the other hand, form gels at much lower solids than high-ester pectins and larger variations of pH are tolerated without major effect on the gel formation. Unlike high-ester pectin, low ester pectin gels may further melt when heated. Amidated low-ester pectins are usually able to jellify preserves, jams, and jellies at the calcium level as is (i.e., with calcium ions originating from fruit and water). Nonamidated low-ester pectins generally require a higher calcium level and addition of ''extra'' calcium is very often necessary to obtain proper gel formation (Figure 4).

Degree of esterification and degree of amidation together control the readiness with which a specific low-ester pectin will react with calcium to form a gel. Nonamidated low-ester

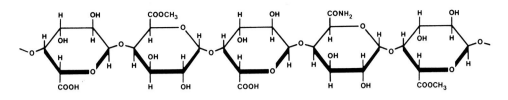

FIGURE 3.   Section of a low-ester pectin molecule with a degree of esterification ~40% and a degree of amidation ~20%.

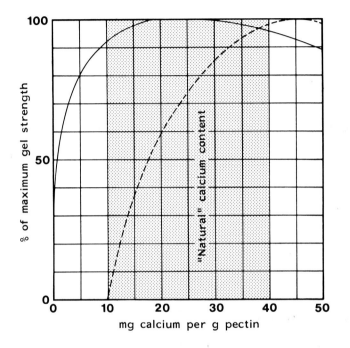

FIGURE 4.   Typical calcium requirement for amidated (—) and non-amidated (---) low-ester pectins. Shaded area indicates range of "natural" calcium usually found in preserves, jams, and jellies.

pectins with a degree of esterification in the range 25 to 35% and amidated low-ester pectins with a degree of esterification in the range 20 to 30% and a degree of amidation of 18 to 25% are usually characterized as highly calcium reactive or rapid setting and find application in systems with a low calcium content or at low soluble solids. Nonamidated low-ester pectins with a degree of esterification in the range 35 to 45% and amidated low-ester pectins with a degree of esterification in the range 30 to 40% and a degree of amidation of 10 to 18% are less calcium reactive, slow set types, and mainly used in systems with a high calcium content or at relatively high soluble solids (Table 1).

Pectins as extracted vary in respect of functional properties and commercial products are therefore most often diluted with sucrose or dextrose for standardization purposes to produce, for example "150 USA-SAG jelly grade" high-ester pectin for use in jams and jellies, or "100 gel power" low-ester pectin for use in products with a reduced sugar content. In addition to sugars, suitable food grade buffer salts may be added for pH control and to achieve desirable setting characteristics.

B. Specifications

Commercial pectins for food applications must generally meet internationally accepted specifications as per the following three publications: (1) FAO Food and Nutrition Paper —

Table 1
MAIN COMMERCIAL PECTIN TYPES

Pectin type	Setting rate designation	Calcium reactivity	Typical degree of esterification (%)	Typical degree of amidation (%)
High ester	Ultra rapid set	Not relevant	76	0
High ester	Rapid set	Not relevant	72	0
High ester	Medium set	Not relevant	68	0
High ester	Slow set	Not relevant	62	0
High ester	Extra slow set	Not relevant	58	0
Low ester	Slow set	Low	40	0
Low ester	Rapid set	High	30	0
Amidated low ester	Slow set	Low	35	15
Amidated low ester	Rapid set	High	30	20

Table 2
OFFICIAL PURITY SPECIFICATIONS FOR COMMERCIAL PECTINS

Reference	14 (FAO)	15 (FCC)	16 (EEC)
Loss on drying (volatile matter)	max 12%	max 12%	max 12%
Acid-insoluble ash	max 1%	max 1%	max 1%
Ash (total)	—	max 10%	—
Sulfur dioxide	max 50 mg/kg	—	max 50 mg/kg
Sodium methyl sulfate	—	max 0.1%	—
Methanol, ethanol, and isopropanol	max 1%	—	max 1%
Nitrogen content, amidated pectin	max 2.5%	—	max 2.5%
Nitrogen content, nonamidated pectin	max 0.5%	—	max 0.5%
Galacturonic acid	min 65%	—	min 65%
Total anhydrogalacturonides in pectin component	—	min 70%	—
Degree of amidation, amidated pectin	max 25%	max 40%	max 25%
Degree of esterification of high-ester pectin component	—	min 50%	—
Degree of esterification of low-ester pectin component	—	max 50%	—
Arsenic, ppm	max 3	max 3	max 3
Lead, ppm	max 10	max 10	max 10
Copper, ppm	max 50	—	—
Zinc, ppm	max 25	—	max 25
Copper + zinc, ppm	—	—	max 50
Heavy metals (as Pb), ppm	—	max 40	—

19,[14] (2) Food Chemicals Codex, 3rd ed.,[15] (3) EEC Council Directive of July 25, 1978.[16] The specifications mainly deal with the identity and purity of the product. A summary of the official purity specifications for pectins intended for use within the food industry is shown schematically in Table 2.

Apart from chemical purity, most pectin manufacturers specify the microbiological purity of their product. As pectin is an acid polysaccharide finding use mainly in acidic media, yeast and mold counts are especially relevant for the user. Typical microbiological specifications may read as follows:

Total plate count (37°C)	Less than 500/g
Yeast and mold count (25°C)	Less than 10/g

E. coli	Test result negative
Salmonella	Test result negative
Staphylococci	Test result negative

Pectins used within the pharmaceutical industry must comply with the specifications for identity and purity in the U.S. Pharmacopeia.[17-19]

C. Standardization

To meet the users' requirement to obtain a pectin with constant properties, most commercial pectins are standardized. High-ester pectins used for jams and jellies are generally standardized to uniform gel strength and setting rate at specified constant conditions. Gel strength standardization is achieved by addition of sucrose or dextrose (ingredients used extensively in all regular jams and jellies) and often expressed as jelly grades.

The "USA-SAG method" based on the work by Cox and Higby[20] and adopted in 1959 by the Institute of Food Technologists[21] is today the most common method for grading high-ester pectins. A jelly grade designation of 150° USA-SAG implies that 1 part of pectin is able to transform 150 parts of sucrose into a jelly prepared under standardized conditions and with standard properties as follows:

1. *Refractometer soluble solids:* 65%
2. *pH:* 2.20 to 2.40.
3. *Gel strength:* 23.5% SAG over 2 min of a gel cast in and removed from a standard glass with exactly specified inner dimensions.

Other methods for grading pectin differ from the USA–SAG method in terms of test jelly preparation, properties, and evaluation. Grade specifications on basis of different methods are not readily comparable or correlated.[22]

The setting rate of high-ester pectins may be standardized according to a method developed by Joseph and Baier.[23] The properties of the test jelly specified are exactly the same as used by the USA-SAG test, and the same jelly batch may in fact be utilized for both determinations. The jelly is exposed to a standard cooling procedure, obtained by placing a USA-SAG standard glass with freshly prepared jelly in a constant temperature (30°C) water bath. The time taken for the jelly to set is measured and typical results obtained for rapid and slow set high-ester pectins are 50 and 225 sec, respectively.

The setting temperature of pectins is sometimes specified and may, in many cases, prove more useful for practical purposes than a setting rate designation, especially if test jelly composition and cooling rate are comparable with actual application conditions. Setting temperature is generally defined as the temperature at which the first sign of jellification can be observed either visually as described by Hinton[24] or by measuring changes in viscoelastic properties or cooling rate.

Gel power of low-ester pectins may be standardized by procedures similar to the USA-SAG method for grading high-ester pectins.[25] However, the idea of standardizing low-ester pectins according to jelly grade has not gained any broad acceptance. Unlike high-ester pectins, low-ester pectins are used in a great variety of products differing in respect of soluble solids content, calcium level, pH, and other factors influencing gel strength performance of the pectin. Market preferences today seem to point towards a variety of low-ester pectin types, each standardized by specific performance tests in systems relevant for the intended application.

III. REGULATORY STATUS

A. Physiological Properties

As a constituent in all land plants, pectin has always been consumed in significant quantities by man. Pectolytic enzymes are found in plants and excreted by many microorganisms. Man, however, possesses no enzyme system to degrade pectin, and it therefore passes unchanged to the large intestine, where bacteria are able to use pectin as their carbohydrate source. Although pectin is hydrolyzed in the intestinal tract, it has no, or only insignificant, net calorie value.[26]

B. Acceptable Daily Intake

Pectins and amidated pectins were evaluated by the "Joint FAO/WHO Expert Committee on Food Additives" (JECFA) in 1981.[27] The committee found that there were no toxicological differences between pectins and amidated pectins, and consequently it was deemed unnecessary to retain the previously established preliminary ADI-value for amidated pectins. A group of ADI "not specified" was established for pectins and amidated pectins, meaning that from a toxicological point of view there are no limitations for the use of pectins, whether amidated or not.

C. U.S. Food and Drug Administration Regulation

Pectins are affirmed "generally recognized as safe" for use in human foods under 21 CFR 184.1558.[28] This affirmation became effective in 1983 and states that pectin is generally recognized as safe for food use when:

1.   It meets the specifications of the *Food Chemicals Codex*, 3rd ed.[15]
2.   It is used as an emulsifier, stabilizer, or thickener.
3.   The levels do not exceed current good manufacturing practice.

In accordance with present practice for "substances whose use has been shown to be safe under reasonably foreseeable conditions of use", the Food and Drug Administration has, besides "good manufacturing practice" not issued any specific limits or guidelines for use levels of pectin in any food.

D. Regulations Outside the U.S.

According to the Canadian Food and Drug Act, the use of pectin is permitted in jams, jellies, and marmalades; in relishes and salad dressings; in certain milk and meat products; and in unstandardized foods at levels only limited by good manufacturing practice. It is further permitted at maximum use level of 0.5% in ice cream products and sour cream and at 0.75% in sherbet.[29]

Food legislation in the member states of the European Economic Community at present distinguishes between pectin (E 440 a) and amidated pectin (E 440 b). Apart from certain limitations regarding the use of amidated pectin in Italy and the Federal Republic of Germany, food regulations in the individual member states generally allow the use of pectin in food products, when a technological need can be proven, at use levels corresponding to "good manufacturing practice." In most other countries, food legislative authorities recognize pectin as a valuable and harmless food additive. If regulated, permitted use levels are generally in accordance with "good manufacturing practice".

IV. MANUFACTURE

A. Raw Material

Apple pomace and especially citrus peel are the only raw materials of any significance for commercial pectin production. Both products are available in ample supply as a residue from juice production and have a high content of pectin with desirable properties, which can be liberated by a relatively simple extraction process. A number of other surplus materials of vegetable origin (e.g., sunflower bottoms and sugar beet waste) have been suggested as basis for pectin manufacture, but apart from a small production of beet pectin in the USSR, none of these alternative sources are utilized today.[30]

Pretreatment of the citrus peel for pectin production involves blanching and washing to terminate pectinesterase activity and remove glucosides, sugars, and citric acid. The peel may further be dried to attain storage stability and allow transportation over longer distances. Dried citrus peel generally contains 20 to 30% pectin.

Apple pomace is normally dried immediately after the juice pressing. Yeast exudates pectolytic enzymes and any fermentation of the fresh apple pomace would rapidly degrade the pectin contained. Dried apple pomace generally yields 10 to 15% pectin.

B. Extraction

To solubilize and liberate the pectin, the raw material — fresh or dried — is added to acidified water (pH usually 1.5 to 3.0). The acid used is most often hydrochloric acid or nitric acid, but sulfuric acid or sulfur dioxide may also come into question.

At the extraction conditions, a certain de-esterification of the pectin will take place, To arrive at the desired degree of esterification at the end of the extraction process, pH, temperature, and time must be controlled carefully. Rapid set high-ester pectins are generally extracted at temperatures close to the boiling point. The high temperature accelerates hydrolysis of the parent pectic substances, lowers viscosity, and facilitates diffusion. Accordingly, the extraction procedure may be finished in less than 1 hr with only minor de-esterification of the pectin taking place. Lower extraction temperature and longer extraction time favors the de-esterification process, yielding slow set high-ester pectins or even low-ester pectins.

C. Purification

The raw extract is a viscous liquid containing 0.3 to 1.5% dissolved pectin and the peel or pomace residue in a more or less swollen and disintegrated state. The separation of the residue from the extract presents one of the key problems in pectin manufacture. Usually, more separation processes, filtrations with various types of filter aid and centrifugations, are necessary to obtain a clear extract. At this stage, the pectin may be further de-esterified by holding at controlled pH and temperature, and the extract may be concentrated by evaporation. Minor quantities of pectin are simply sold as the concentrated extract. This ''liquid pectin'' is normally preserved with sulfur dioxide and delivered in barrels or by tank car.

D. Isolation

The pectin may be isolated in its pure form either by precipitation with alcohol or by precipitation as an insoluble salt by addition of suitable cations, usually aluminum. A simple drum drying of the concentrated pectin extract leads to a commercial product containing all extracted impurities and is hardly used anywhere today.

Alcohol precipitation is obtained by mixing the extract with methanol, ethanol, or 2-propanol. The gelatinous precipitate obtained is washed with alcohol and pressed to remove soluble impurities, and finally dried and milled to yield powdered pectin.

Aluminum precipitation involves addition of a concentrated solution of either aluminum sulfate or aluminum chloride, together with alkali (for pH-control) to the purified extract. By this procedure, pectin is isolated as a co-precipitate with aluminum hydroxide. To obtain a pure pectin, it is necessary first to wash the precipitate with acidified alcohol to remove aluminum, and second, to wash with alkaline alcohol to neutralize the product. Pressing, drying, and milling to obtain a powdered pectin finalize the process. The various production methods and production steps in high-ester pectin manufacture are shown in a condensed form in Figure 5.

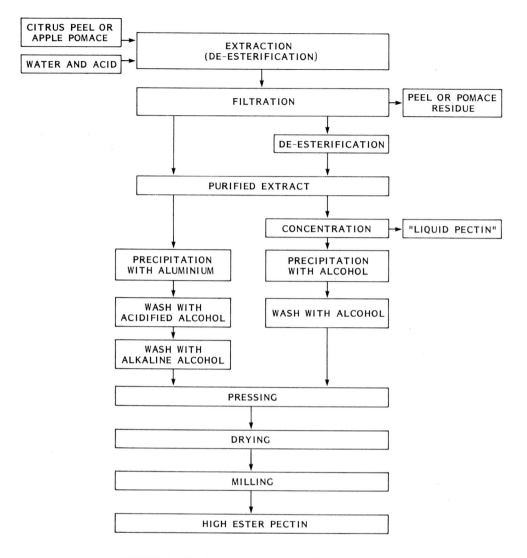

FIGURE 5.   Principal production methods for high-ester pectins.

E. De-Esterification

The de-esterification of the extracted pectin required to produce low-ester pectin types may, as explained before, take place during extraction or in the purified extract. However, to avoid undesired depolymerization, it is often preferred to postpone the de-esterification till the pectin has been precipitated. The precipitate is dispersed in alcohol and either acid or alkali is added to obtain the desired reaction. If the base used is ammonia, a certain amidation in addition to the de-esterification will take place.

The use of specific microbial pectin esterases for the production of low-ester pectins has been suggested.[31] Unlike pectin esterases of vegetable origin, certain microbial pectin esterases will lead to a random distribution pattern of free carboxyl groups, similar to that obtained by acid- or alkali-induced de-esterification.[32] The idea is, however, hardly economic for any industrial utilization at present.

Flow diagrams showing various production methods for low-ester pectins are presented in Figure 6.

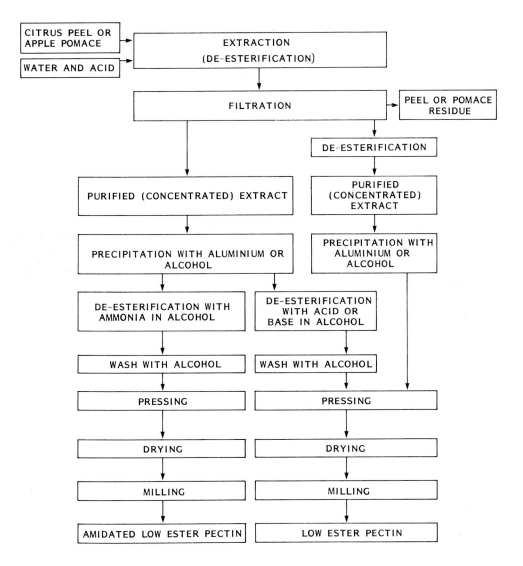

FIGURE 6. Principal production methods for low-ester pectins.

F. Pectin Manufacturers

Pectin production originally sprang up in citrus- or apple-growing and processing areas, where fresh raw material was abundant. Today, however, a number of major producers have grown to a size where production, wholly or partly, must be based on imported dried citrus peel.

Hercules, Inc., Wilmington, Del., is the largest pectin producer at present, owning pectin factories in Denmark (The Copenhagen Pectin Factory Ltd.), Germany (Pomosin AG), Italy (Cesalpinia S.P.A.), and since 1980 also producing domestically in Florida. Other large producers are Unipectine S.A. in France, HP Bulmer Limited in England, Pektin-Fabrik Hermann Herbstreith KG in Germany, Atlantic Gelatin Plant, General Foods Corp., Massachusetts, and Pectina de Mexico S.A. (owned by Grindsted Products Ltd.) in Mexico. Smaller pectin manufacturers are found in Switzerland, Brazil, Israel, Argentina, and in a number of East European countries.

V. STRUCTURE

A. Conformation and Structural Randomness

As a complex heteropolysaccharide with varying composition according to origin and extraction conditions, pectin cannot be characterized in terms of a specific overall structure and conformation. It is, however, possible to recognize distinctive structural elements contained in all pectin material.

Functional properties, especially gel formation and rheological phenomena, involving interactions between more molecules, depend on the presence of specific structural regions in the pectin molecular chains. Small variations in composition may cause spatial and conformative changes which totally invalidates the functional properties of the product. Pectins extracted from sugar beets and potatoes contain a high proportion of galacturonic acid units in which the hydroxyl groups at position C-2 and C-3 are esterified with acetate. As a result, sugar beet and potato pectin do not form gels at all, unless the acetate groups have been removed chemically or enzymatically. Depending on origin, extraction procedure, and preparation, hydrolysates of pectin always contain varying amounts of neutral sugars, especially D-galactose, L-rhamnose, and L-arabinose.[33,34] The neutral sugars are to a large extent constituents in side chains to the polygalacturonan backbone, a remainder from the complex protopectin structure, but 1,2-linked L-rhamnose will also be present in the main polygalacturonic chain. Rhamnose insertions will provide "kinks" in the molecular chain, a basic structural condition of a molecular gel network formation.[35]

A comprehensive understanding of pectin structures and the consequences regarding conformative and interactive aspects has, by far, not been reached yet. Prevailing models concerning fundamental structure elements are even contradictive in certain ways. Different findings may to a large extent reflect the heterogeneity of the product and stem from variations in raw material and preparative methods. X-ray diffraction studies on dried fibers indicate that the galacturonan backbone forms a right-handed helix, with 3 galacturonic acid units in C_1 conformation as the repeating sequence, corresponding to a repeat distance of 13.4 Å.[36-38] The 3_1 helical structure is stabilized by intramolecular hydrogen bonds to form a fairly stiff rod-like structure, which may be interrupted by irregularities in the molecule as, for example, rhamnose insertions (Figure 7). From circular dichroism and equilibrium dialysis investigations concerning calcium binding to poly-D-galacturonate and low ester pectins, Morris et al.[39,40] have suggested that gel formation with calcium involves polygalacturonic acid sequences with a 2_1, ribbon-like symmetry. On drying the gel, however, the 3_1 helical symmetry is restored through a polymorphic phase transition.

Distribution of free carboxyl groups along the pectin molecules and the distribution of molecules with different degree of esterification within a specific pectin preparation has been investigated by various techniques.[41-48] Enzymatic de-esterification in vivo tends to produce pectins with free carboxyl groups occurring in blocks and a large variation in degree of esterification among the contained pectin molecules. De-esterification in vitro with acid or alkali leads to a random distribution of free carboxyl groups and relatively small variations in degree of esterification between the individual pectin molecules. Accordingly, commercial pectins normally show a free carboxyl group distribution pattern in between the two, determined by extent of in vivo de-esterification in the raw material and subsequent in vitro de-esterification during the manufacturing process.

Compared with pectin preparations extracted at mild conditions, commercial pectins generally contain lower quantities of neutral sugars. This can be explained as a result of acid hydrolysis of neutral sugar side chains during the industrial extraction process. Accordingly, a large part of the neutral sugars left in commercial pectins is 1,2 bound rhamnose present in the galacturonan backbone.

Distribution of rhamnose insertions along the pectin molecular chains has not been fully elucidated at present. Partial acid hydrolysis of pectins of various origin tend to produce

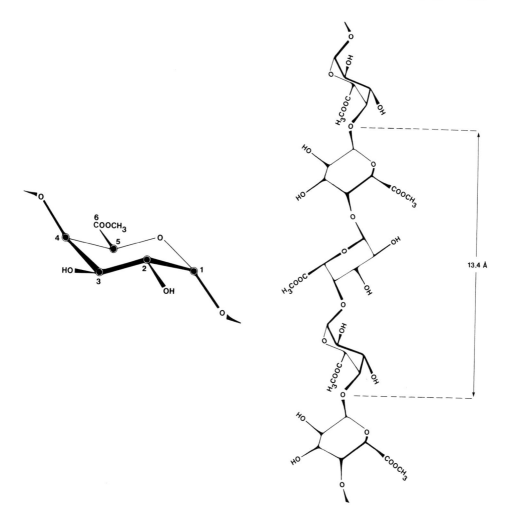

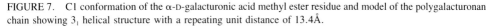

FIGURE 7.   C1 conformation of the α-D-galacturonic acid methyl ester residue and model of the polygalacturonan chain showing 3_1 helical structure with a repeating unit distance of 13.4Å.

galacturonan segments of fairly constant size corresponding to 25 units. As the glucosidic bond betwen C_1 in rhamnose and C_4 in galacturonic acid is considered less acid stable than other glucosidic bonds in the molecule, Powell et al.[40] have suggested that the length of polygalacturonate sequences between rhamnose interruptions is fairly constant and corresponding to approximately 25 residues. (Figure 8 A.) Neukom et al.[49] have analyzed similar sequences with a degree of polymerization of 20 to 30, obtained by acid degradation of apple pectin preparations. Their experiment showed that the oligomers were almost fully made up by galacturonic acid units and only traces of rhamnose were present. They were, on the other hand, also able to isolate galacturonan segments containing rhamnose from apple tissue and concluded that the cell walls contained both a pure galacturonan-type pectin and a rhamnogalacturonan-type pectin, possibly located in different regions.

Based on results obtained by specific enzymatic degradation of apple pectins, De Vries et al.[50,51] have suggested a molecular model consisting of a long homogalacturonan chain intercepted by a few relatively short "hairy" regions containing all rhamnose insertions and side chains (Figure 8B).

B. Molecular Weight

Molecular weight of pectins can be expected to vary with raw material, extraction con-

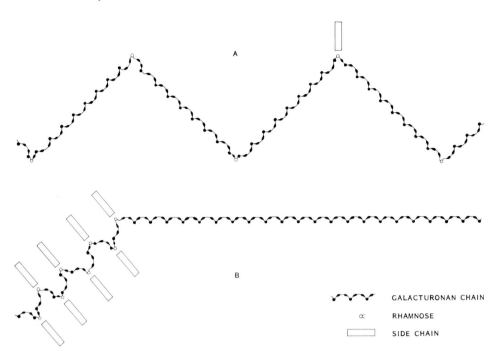

FIGURE 8.   Contemporary models for occurrence of side chains and rhamnose insertions in the pectin molecule. (A) Even distribution, as suggested by Powell et al.[40] (B) Blockwise occurrence in a few hairy regions, as suggested by De Vries et al.[50,51]

ditions, and isolation procedure. Based on viscosimetry, the average molecular weight of commercial pectin normally falls within the range 50,000 to 150,000 daltons.[42,52,53]

Other techniques used for molecular weight determination (e.g., light scattering) often result in apparent molecular weight around 1 million or even higher. The reason for this inconsistency is a considerable amount of intermolecular association even in fairly dilute solutions leading to aggregates of pectin molecules.[54–57]

VI. PROPERTIES

A. Powder Properties and Solubility

In the manufacturing process, pectins are normally dried to a water content below 10%. As the equilibrium moisture content is 12% in 70% relative humidity, pectin tends to take up moisture in most climates. To avoid caking and ensure optimum storage stability, it is recommended to keep the product in a vapor-tight package at cool and dry conditions.[58]

High-ester pectins will slowly loose gel power and gradually adopt slower setting characteristics due to depolymerization and de-esterification during storage. At a storage temperature of 20°C, a loss of gel power of 5%/year must be expected; at higher storage temperatures, the change in functional properties will take place at a much faster rate. Low-ester pectins are considerably more stable than high-ester pectins. When the product is kept at room temperature, it is scarcely possible to detect any loss of functional properties over 1 year.

Commercial pectins are typically milled to a particle size corresponding to a 99% pass through a 60 mesh (0.25 mm) sieve. The powder has usually a fairly low density, around 0.7 g/cm³ and a somewhat fibrous structure with powder flow properties inferior to a crystalline product. Pectin is generally soluble in water and insoluble in most organic solvents. Water solubility increases with decreasing molecular weight and increasing degree of randomness of the carboxyl group distribution. Sodium pectinates are generally more soluble than pectinic acids, which in turn are more soluble than calcium pectinates. Pectic acids and pectinic acids with a low degree of esterification are only soluble as the sodium or potassium salt. Increasing amounts of sugar or calcium ions in the water will make the pectin more and more difficult to dissolve. It is accordingly recommended always to dissolve pectin at soluble solids below 25% and preferably in hot water to ensure maximum hydration and disintegration of the large molecular aggregates present in the pectin powder.

When added to water, powdered pectin tends to form lumps consisting of pockets of semidry powder enrobed by a shell of hydrated material. These lumps dissolve very slowly, and a rational dissolution process accordingly involves suitable techniques to obtain a perfect dispersion of the powder. Vigorous agitation as obtained with a high-speed mixer may serve to disperse the pectin before any lumps are formed and subsequently dissolve the pectin completely within a few minutes. When hot (60 to 90°C) water is used, up to 10% pectin solutions may be made using this technique. However, to avoid handling problems due to too high viscosity, only 5 to 8% solutions are normally produced. Complete dispersion may be obtained with conventional stirrers, if the pectin is dry blended with approximately 5 parts of inert material (most often sucrose) or initally dispersed in media where pectin is insoluble (e.g., alcohol or a concentrated sugar syrup). Dry blending with 2% sodium bicarbonate will further facilitate dispersion as carbon dioxide, developed during the dissolution process by the reaction between bicarbonate and the acidic pectin, will tend to break up any lumps formed.

B. Solution Properties

Pectin solutions show relatively low viscosities compared to other plant hydrocolloids, and pectin consequently has limited use as a thickener. The viscosity increases with increasing degree of polymerization and increasing degree of esterification, but is further affected by the presence of various solutes in the system (Figure 9). Basically, the viscosity of a pectin solution will depend on size and shape of the molecules, but formation of macromolecular aggregates and interactions with electrolytes or water molecules can be expected to affect the flow properties of the solution.

In dilute solutions, where interactions between more pectin molecules are negligible, viscosity increases with increasing dissociation of the carboxyl groups and reaches a plateau when almost full dissociation has been obtained.[59] This phenomenon may be interpreted as an increase of the hydrodynamic volume of the molecule caused by repulsion between adjacent charged carboxyl groups. The pK value of pectin is approximately 3.3, and the observed viscosity increase corresponds to a change in pH from 2 to approximately 5.[60,61] The intramolecular repellance of charged carboxyl groups may be counteracted by an increase in ionic strength of the solution. Addition of sodium chloride to a pectin solution has accordingly been shown to lead to a remarkable drop in viscosity.[10,59] Viscosity of high-ester pectin solutions generally increases with increasing degree of esterification.[62] This effect may be explained by an increase in the structuring of water molecules around the methyl ester groups.[38] In the case of low-ester pectins, low degrees of esterification or a blockwise occurrence of carboxyl groups will lead to high viscosities reflecting an inferior solubility of these pectins.

Dilute pectin solutions show near to Newtonian flow properties, but these change to pseudoplastic as the pectin concentration is increased. This change in rheological properties indicates an interaction between pectin molecules in more concentrated solutions resulting in an integral structure which can be broken mechanically, but reforms when the shearing action is stopped.

Addition of calcium or other divalent cations to pectin solutions leads to the formation of strong intermolecular chelate bonds involving carboxyl groups occurring in sequences of

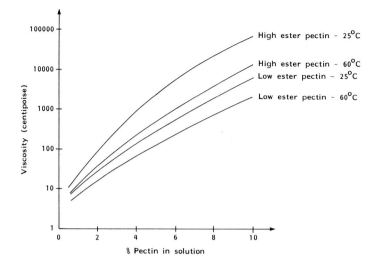

FIGURE 9.   Typical viscosity of commercial high-ester and low-ester
pectin at 25 and 60°C.

suitable length and the metal cation in question.[63] Addition of calcium to low-ester pectin
solutions may even cause formation of a true gel which will not reform after mechanical
rupture.

Intermolecular interactions leading to aggregation of high-ester pectin molecules are favored by decreasing the charge on the molecules (i.e., by lowering the pH) and by addition
of cosolutes, normally sucrose or other carbohydrate sweeteners. Addition of water-soluble
alcohols and ketones to pectin solutions causes precipitation of the polymer. Pectin is further
precipitated by heavy metals, by quaternary detergents, and by most cationic macromole-
cules.[64,65] High-ester pectin will interact with casein particles in sour milk products to yield
a stablilized suspension of casein particles which can be pasteurized without coagulation or
precipitation.[66]

Application procedures involving holding of pectin solutions at elevated temperatures
leads to depolymerization and de-esterification of the pectin. Both processes are mainly
governed by the. pH of the solution. Compared to other gelling agents, pectin shows a
remarkable stability in the pH range 3 to 4.5. At lower pH values and elevated temperatures,
degradation due to hydrolysis of glucosidic bonds is observed. De-esterification is also
favored by low pH, and long holding times may cause a high-ester pectin gradually to adopt
slower setting characteristics.

At pH values above 4.5, high-ester pectin is stable only at room temperature. At elevated
temperatures, chain cleavage by β-elimination rapidly leads to loss of viscosity and gelling
properties (Figure 10). Only glucosidic bonds next to an esterified carboxyl group can be
broken by β-elimination, and low-ester pectins are accordingly considerably more stable at
higher pH values than high-ester pectins. This phenomenon explains, for example, why it
is possible to UHTST-sterilize milk desserts with low-ester pectin as the gelling agent.

C. Gel Formation of High-Ester Pectin

Any solution containing high-ester pectin at potential gelling conditions has an upper
temperature limit above which gelation will never occur. When cooling the solution below
this temperature, gel formation takes place after a certain time. Practical experience shows
that the observed gelling temperature and gelling rate mainly depend on the following factors:

1.   *Degree of esterification* — Increase in degree of esterification leads to higher setting
     temperatures and increased setting rate.

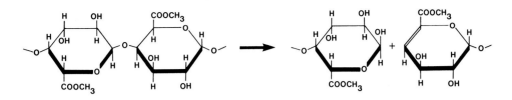

FIGURE 10. β-elimination.

2. *pH* — Reduction of pH leads to increase in gelling temperature and setting rate. Commercial high-ester pectins will generally not form gels at pH values above 4.0, and normal application conditions usually require a pH around 3.0 for proper gel formation.

3. *Soluble solids* — Increase in sugar content leads to increase in gelling temperature and setting rate. Practical working range for commercial high-ester pectins is between 55 and 80% soluble solids. Below 55% soluble solids, no gel formation is obtained, and at solids above 80%, setting will occur practically at the boiling point of the system.

Polysaccharide gel formation generally involves crosslinking of the polymer to form a three-dimensional network in which water, sugar, and other solutes are held. A model for the junction zones in the high-ester pectin gel network has been suggested by Walkinshaw et al.[38] According to this model, 3 to 10 polymer chain segments with a 3_1 helical structure will form aggregates of parallel chains limited in size because of steric barriers, entropic factors and possible rhamnose insertions. This local crystallization would be sustained by intermolecular hydrogen bonds and probably reinforced by hydrogen bonding with water molecules in one set of triangular channels and hydrophobic attractions between methyl groups forming columns in a second set of triangular channels. (See Figure 11.)

Gel formation of high-ester pectin can accordingly be explained as local polymer aggregations made possible by the combined effect of cosolutes (sugars) breaking up the cages of water molecules surrounding the individual polymer chains, a low pH resulting in protonation of carboxyl groups and hence a decrease in electrostatic repulsion between the galacturonan chains, and finally the presence of methyl ester groups lowering total charge of the molecules and actively contributing to the chain interactions.[68,69]

Rheological properties of high-ester pectin gels indicate that at least two types of bonding are involved in the molecular gel network.[70] One type of bond is strong and responsible for the elastic properties of the gel, while the other type of bond is weaker and capable of reforming after disruption. When sucrose as cosolute is substituted by corn syrup, a remarkable increase in gelling temperature will occur. This observation indicates that sugars play an active role in the formation of the pectin gel network. Most likely pectin molecules associate with sugar molecules through hydrogen bonds to form secondary links reinforcing the molecular network structure.

Gel strength of high-ester pectin gels generally increases with increasing pectin concentration and increasing molecular weight of the pectin. Homogeneity and specific structural characteristics of the pectin material in question plus the preparative method used to produce the gel will further influence the gel strength obtained. High-ester pectin gels are generally cohesive and show no tendency to syneresis. The gels are not temperature reversible, but will slowly dissolve in hot water.

D. Gel Formation of Low-Ester Pectin

Low-ester pectins form gels in the presence of calcium and other divalent metal ions. Setting temperature of low-ester pectin gels generally increases with increasing de-esteri-

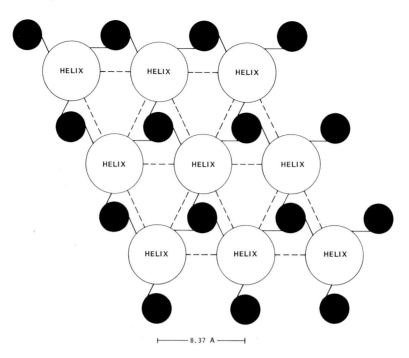

FIGURE 11.   Model of junction zone in a high-ester pectin gel as suggested by Walkinshaw et al.[38] Parallel polygalacturonan chains, viewed along the helix axis, are packed in a hexagonal lattice. The structure is sustained by intermolecular hydrogen bonds (dotted lines) and reinforced by columns of methyl groups (filled circles).

fication of the pectin, increasing amount of calcium and soluble solids in the system, and increasing amount of possible galacturonamide substitutions in the molecule.

Kohn et al.[63,71,72] have investigated the binding of calcium ions to oligogalacturonates, polygalacturonates, and pectinates. The investigations proved that calcium binding to low-ester pectin can not be explained as a simple electrostatic interaction, but involves intermolecular chelate binding of the cation leading to the formation of macromolecular aggregates. The results further indicated that sequences of minimum 10 galacturonic acid units in the polymer are required to obtain intermolecular association with calcium ions.

Rees[73,74] has suggested a so-called "egg-box" model for the primary junction zones in the low-ester pectin molecular gel network. Chain segments of 14 or more residues with a 2_1 ribbonlike symmetry are believed pairwise to form parallel oriented aggregates. Calcium ions forming chelate bonds with oxygen atoms from both galacturonan chains will fit into "cavities" in the structure. (See Figure 12.)

The extension of the crystallites will be limited by methyl ester residues and by rhamnose insertions side chains or other irregularities in the chains. Secondary junction zones in the gel network may arise from hydrogen bonding with water and sugar molecules and, if calcium is present in surplus amount, from subsequent calcium induced aggregation of the preformed dimers. The presence of galacturonamide substitutions in the low-ester pectin molecular chain is known to promote gel formation but also to counteract network collapse at higher calcium levels. (See Figure 4.) However, the mechanisms responsible for these modifications of the gel properties have not been elucidated yet.

Gel strength of low-ester pectin gels varies with pectin concentration, pH, and calcium content. Optimum gel strength is normally found in the pH range 3.0 to 3.5, but excellent gels may be obtained at higher pH values by increasing pectin dosage or calcium level. Gel strength increases with increasing calcium content, but higher calcium levels lead to brittle

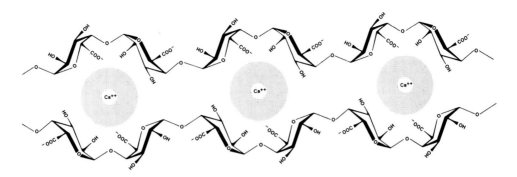

FIGURE 12. "Egg-box" model of a primary junction zone in a low-ester pectin gel.

gels with syneresis tendency and eventually to disintegration of the structure by extensive crystallization. Optimum calcium content required to produce coherent gels with minimum syneresis tendency depends primarily on degree of esterification of the pectin in question. Quality of commercial low-ester pectins is accordingly closely connected to inter- and intramolecular consistency of the material.

Increasing amounts of soluble solids lead to less brittle gel structures and reduced syneresis tendency, but further increase setting temperature of the gel. The increase in setting temperature may be counteracted by selecting low-ester pectin types with a higher degree of esterification or by decreasing the calcium content of the system. Addition of polyphosphates or other calcium complexing agents will also decrease the gelling temperature. Unlike high-ester pectin gels, low-ester pectin gels weaken with increasing temperature and will usually melt when heated to a temperature 5 to 10°C above the setting temperature.

VII. FOOD APPLICATIONS

A. Function in Foods

Pectin is primarily used as a gelling agent in foods. Dependent on pectin type and dosage, and the composition of the food system, textures ranging from soft and thixotropic to firm cohesive or brittle gels can be obtained. Gel formation is most often obtained by cooling the product below the gelling temperature of the system. However, the specific gelling mechanism of pectin also offers the possibility to obtain cold gelation by addition of acid, sugar, or calcium to a food system containing pectin in solution.

Gel formation of pectin may be utilized to stabilize multiphase foods, either in the final product or at an intermediate stage in the process. The anionic character of the polymer may, in specific systems, also serve to counteract aggregation of, for example, protein particles. The thickening effect of pectin in terms of pure viscosity increase is utilized mainly where food regulations prevent the use of cheaper gums or where an "all natural" image of a product is essential.

The main food applications of pectin are listed in Table 3. Close to 80% of all pectin produced is consumed by the fruit processing industry and the group, jams, jellies, and preserves, is by far the most important application area.

For detailed information about formulations and application procedures for the various foods listed, it is recommended to consult commercial handbooks and technical data sheets as published by the pectin manufacturers.[75-78]

B. Jams, Jellies, and Preserves

Pectin is used to impart a texture to jams, jellies, and preserves that allows transportation without changes, gives a good flavor release, and minimizes syneresis. During the manu-

Table 3
FOOD APPLICATIONS OF PECTIN

Product group	Function of pectin	Pectin use level (%)	Estimated consumption (metric tons/year)
Jams, jellies, and preserves	Gelling agent, thickener	0.1—1.0	10,500
Bakery fillings and glazings	Gelling agent, thickener	0.5—1.5	500
Fruit preparations	Thickener, stabilizer	0.1—1.0	400
Fruit beverages and sauces	Thickener, stabilizer	0.01—0.5	400
Confectionery	Gelling agent, thickener	0.5—2.5	2,000
Dairy products	Stabilizer, gelling agent	0.1—1.0	1,100
Miscellaneous	—	—	100

facture of a jam, the pectin must ensure uniform distribution of fruit particles in the continuous jelly phase from the moment the mechanical stirring ceases, i.e., the pectin must set quickly after the filling operation.

Regular jams with a refractometer soluble solids content between 65 and 70% are usually made with high-ester pectins. Rapid set pectins are used for products filled into smaller jars at temperatures around 90°C. When filling into larger packages, the filling temperature must be reduced to avoid heat damages of the fruit, as product in the middle of the large pack only cools very slowly, irrespective of outside temperature. To avoid premature gel formation of the pectin, slow set types must accordingly be used for products in larger packages. Slow set pectins are always used for jellies, as the relatively low setting temperature allows ample time for any air bubbles to escape from the product before gel formation starts (Figure 13).

Low-ester pectins may be used in jams with solids ranging from 25 to 75%. However, in regular jams and jellies, low-ester pectins are less economic in use than high-ester pectins and generally only used if a soft, spreadable and thixotropic texture is desired. Low sugar jams and fruit spreads are, on the other hand, generally produced with low-ester pectins. Calcium-reactive types find use at lower solids while types with a relatively higher degree of esterification are used at solids around 55%. At soluble solids below 25%, reactive low-ester pectins may yield products with an adequate gel texture. When the gel is broken, however, syneresis will cause considerable juice exudation. At low solids, low-ester pectins are accordingly often used in combination with other water-binding hydrocolloids.

The amount of added pectin to be used in a jam or jelly depends on desired consistency, amount of added fruit, pectin, or calcium content in the fruit used, soluble solids, heat treatment, and packing size. To obtain a jellified yet spreadable texture of a jam containing 50% fruit and with 65% soluble solids typically 0.2 to 0.3% high-ester pectin is required. Approximate changes necessary in the pectin concentration in a jam formula when soluble solids are reduced or increased, the consistency remaining unchanged, can be calculated from Table 4.

Apart from the industrial use of pectin for jams and jellies, a considerable amount of high- and low-ester pectins marketed as gelling powders or gelling liquids is used for homemade jams and jellies.[79]

C. Bakery Fillings and Glazings

Fruit-based products are important ingredients within the bakery industry, their acidulous taste and relatively high moisture content contrasting to the baked goods.

Oven-resistant high sugar jams may be produced with high-ester pectins, if the recipe used ensures high temperature stability of the product. This is obtained if the solids content is kept relatively high (i.e., around 70%) and by using a relatively rapid setting pectin type. To obtain a satisfactory result, however, it is necessary that the product is not extensively ruptured mechanically (by stirring or pumping). Mechanical rupture of the gel tends to

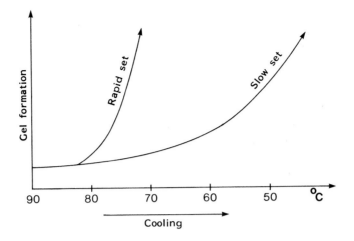

FIGURE 13.   Typical gel formation in jams produced with rapid set and slow set high-ester pectin.

Table 4
PECTIN TYPE AND RELATIVE CONCENTRATION RECOMMENDED FOR JAMS AND JELLIES WITH VARIOUS SOLUBLE SOLIDS CONTENT

Percent soluble solids in jam	Recommended pectin type	Relative pectin concentration required
75	Slow-set high-ester pectin	77
70	Slow set high-ester pectin	88
65	Rapid set high-ester pectin Slow set high-ester pectin	100 (basis)
60	Rapid set high-ester pectin	117
55	Moderate calcium-reactive (slow set) low-ester pectin	135
45	Moderate calcium-reactive (slow set) low-ester pectin	145
35	Calcium-reactive (rapid set) low-ester pectin	175
25	Calcium-reactive (rapid set) low-ester pectin	210

initiate syneresis which is greatly increased when heating to oven temperatures. By using nonamidated low-ester pectins, it is possible to produce bakery jams with satisfactory stability at oven temperatures and minimum syneresis tendency after breakage of the gel texture by pumping or stirring. An optimum result, however, requires a relatively high pectin dosage (0.7 to 1.0%) and a strict control of calcium content in the system.

A number of pectin-based products are used for decoration purposes within the bakery industry. As these fillings and glazings normally are applied to the baked good after the baking, heat stability is of no importance. Semimanufactured liquid jelly bases containing pectin in solution are often used for industrial bakery lines. Cold gelation is obtained by a continuous addition of acid or calcium to the liquid base immediately before the filling or glazing operation.

The heat reversibility of low-ester pectin gels may also be utilized in bakery glazings. By formulating a jelly base with a soluble solids of approximately 65% and a relatively high dosage of a calcium reactive low-ester pectin, a pregelled product with a paste-like texture and a relatively good microbiological stability is obtained. Prior to application in the bakery, water is added to the jelly base, and the diluted product is heated to 85°C to melt the low-ester pectin gel. The product may now be used for "hot glazing" of fruit tarts and similar products. By cooling, the low-ester pectin gels at optimum conditions — due to the dilution with water — to form a coherent and glossy jelly.

D. Fruit Preparations

Industrial fruit preparations are mainly used for combination with yogurt and other dairy products. The product is often aseptically filled into large containers (e.g., 1000-kg containers) at relatively low temperatures. Low-ester pectin is usually incorporated into fruit preparations to create a soft, thixotropic gel texture, sufficiently firm to ensure uniform fruit distribution, but still allowing the product to flow out of the container by gravity and endure pumping without disintegration of the texture.[76,80]

E. Fruit Beverages and Sauces

High-ester pectin is used in fruit drink concentrates to stabilize oil emulsion and fruit particle suspension. In this application, the gelation is apparent in the end product only as a thickening effect, as the coherent gel texture[83] has been broken mechanically to obtain a smooth flow. Pasteurization or extensive homogenization must not be used, however, as these treatments tend to change the rheological properties of the system to an extent where the stabilizing effect is lost.

The stabilizing and mouthfeel creating properties of pectin are utilized in recombinated juice products and, especially artificially sweetened, soft drinks.[81,82] Pectin is further incorporated into instant fruit drink powders to produce a natural mouthfeel in the drink.

Low-ester pectins, alone or in combination with other food hydrocolloids, are used in fruit or tomato sauces, toppings, table syrups, and ripples to create a thickened, semigelled texture.

F. Confectionery Products

Slow set high-ester pectins are mainly used within the confectionery industry for making fruit jellies and jelly centers, flavored with natural fruit constituents or synthetic flavors. In combination with whipping agents, pectin is further used as a texturizer for aerated fruit-flavored products.

Buffered low-ester pectins not requiring addition of acid for gel formation are used for jellies and centers in which the low pH range necessary for high ester pectin gelation is not acceptable for flavor reasons (e.g., peppermint- or cinnamon-flavored jelly beans). At low concentrations, low-ester pectin may further impart a thixotropic texture to confectionery fillings. At higher concentrations, a cold gelation can be obtained if calcium ions are allowed to diffuse into the product.[77,84,85]

Compared to other gelling agents commonly used for confectionery products, pectin requires that recipe and production parameters are strictly observed, but offers the advantage of an excellent texture and mouthfeel, an extremely good flavor release, and compatibility with modern continuous processing due to a fast and controllable gelation. Commercial pectins manufactured to yield specific solubility characteristics may dissolve in systems with a high soluble solids (i.e. 75 to 80%), if heated under pressure to a temperature close to 140°C. Such pectins are used for confectionery jellies or centers produced continuously on jet-cooking lines or with pressure dissolvers, equipment extensively used within the confectionery industry for a variety of products.

G. Dairy Products

The "protective colloid" effect of high-ester pectin is utilized to stabilize sour milk drinks either cultured or produced by direct acidification (fruit juice-milk combinations).[66,78,86,87]

The pectin reacts with the casein, prevents the coagulation of the casein at pH below the isoelectric pH (4.6), and allows pasteurization of the sour milk products to extend their shelf life. The stabilizing effect is not limited to casein particles. It is, for example, possible to produce stable acidified whey drinks and soy milk drinks with pectin.

The texture of yogurt may be improved by small amounts of low-ester pectin added to the milk before pasteurization and fermentation. Low-ester pectins also form gels with calcium present in milk at higher pH. Low-ester pectin is hence suited as gelling agent for milk desserts, but less economic in use than, for example, carrageenan, which gels milk at a much lower use concentration. Low-ester pectin may, however, be preferred as gelling agent for milk desserts combined with acidic fruit sauces. Unlike carrageenan, low-ester pectin will not react with casein in the gel to form a cheeselike precipitate at the interface where pH is reduced by hydrogen ion diffusion.

The calcium reactivity of low-ester pectin may be utilized when adding milk (calcium ions) to a fruit syrup containing low-ester pectin. A canned fruit syrup with 20 to 30% soluble solids and pH 4.0, containing 2% low-ester pectin with moderate calcium reactivity in solution will, when mixed with an equal amount of cold milk, quickly produce a fruit-flavored gelled milk dessert.

H. Miscellaneous

The use of low-ester pectin as gelling agent and texturizer has been suggested in numerous food products, ranging from artificial caviar and meat products to dessert jellies.[88-90]

A synergistic effect between pectin and alginate regarding gel formation properties has been reported.[69,91] In combination with xanthan gum, pectin is used as a stabilizer for salad dressing.[92,93] Incorporation of pectin in water ice and sherbet reduces ice crystal growth and improves mouthfeel and melting properties.[94] Low-ester pectins may, in higher concentrations, be used to produce a frozen water gel on a stick. In combination with galactomannans or carboxymethylcellulose, pectin is further used in ice cream stabilizers.[95]

Pectins and reaction products with pectin have been suggested as emulsifiers for low-calorie mayonnaise and fruit butters.[96] Optimum effect in forming and stabilizing emulsions has been reported to depend on pH of the system.[97] The quality of frozen fruit may be improved by incorporation of low-ester pectin, adding firmness to the product and reducing juice exudation.[98] Shelf life and appearance of dehydrated or candied fruits may be improved by coating with an edible low-ester pectin film.[99] Binding and texturizing properties of pectin have been reported to influence the physical properties of spray-dried instant tea and dietetic bread positively.[100,101] As a curiosity, it can finally be mentioned that pectin has found its way as binder and texturizer into the standard "space menu" of Russian astronauts.[102].

REFERENCES

1. **Vauquelin, M.,** Analyse du tamarin, *Ann. Chim. (Paris),* 5, 92, 1790.
2. **Braconnot, M. H.,** Recherches sur un novel acide universellement répendu dans tous les végétaux, *Ann. Chim. Phys.,* Ser. 2, 28, 173, 1825.
3. **Kertesz, Z. I.,** *The Pectic Substances,* Interscience, New York, 1951.
4. **Deuel, H. and Stutz, E.,** Pectic substances and pectic enzymes, *Adv. Enzymol. Relat. Subj. Biochem.,* 20, 341, 1958.
5. **Doesburg, J. J.,** Pectic substances in fresh and preserved fruits and vegetables, I.B.V.T. Commun. No. 25, Institute for Research on Storage and Processing of Horticultural Produce, Wageningen, 1965.
6. **Glicksman, M.,** *Gum Technology in the Food Industry,* Academic Press, New York, 1969, 159.
7. **Pilnik, W., Zwiker, P., Pektine,** *Gordian,* 70, 202—204, 252—257, 302—305, 343—346, 1970.
8. **Christensen, O. and Towle, G. A.,** Pectin, in *Industrial Gums,* 2nd ed., Whistler, R. L., Ed., Academic Press, New York, 1973, 429.

9. **Petit, R.,** Les matières pectiques, *Afinidad,* 32, 585, 1975.

10. **Kawabata, A.,** Studies on chemical and physical properties of pectic substances from fruits, in *Memoirs of the Tokyo University of Agriculture,* Vol. 19, The Tokyo University of Agrigulture, Tokyo, 1977, 115.

11. **Pedersen, J. K.,** Pectins, in *Handbook of Water-Soluble Gums and Resins,* Davidson, R. L., Ed., McGraw-Hill, New York, 1980, chap. 15.

12. **Pilnik, W. and Voragen, A. G. J.,** Pektine und Alginate in Gelier- und Verdickungsmittel in Lebensmitteln, Neukom, H. and Pilnik, W., Eds., Forster Verlag AG, Zürich, 1980, 67.

13. **Baker, G. L., Joseph, G. H., Kertesz, Z. I., Mottern, H. H., and Olsen, A. G.,** Revised nomenclature of the pectic substances, *Chem. Eng. News,* 22, 105, 1944.

14. **Anon.,** Amidated pectin, Pectins, in Specifications for identity and purity of carrier solvents, emulsifiers and stabilizers, enzyme preparations, flavouring agents, food colours, sweetening agents and other food additives, FAO Food and Nutrition Paper 19, Food and Agriculture Organization of the United Nations, Rome, 1981, 10—14, 152—155.

15. **Anon.,** Pectin, in *Food Chemicals Codex,* 3rd ed., National Academy Press, Washington, D.C., 1981, 215.

16. **Anon.,** E 440(a) — Pectin, E 440(b) — Amidated pectin, in Council Directive of July 1978 laying down specific criteria of purity for emulsifiers, stabilizers, thickeners and gelling agents for use in foodstuffs, Official Journal of the European Communities, L 223, 1978, 16.

17. **Anon.,** Pectin, in *The United States Pharmacopeia,* 20th rev., United States Pharmacopeial Convention, Rockville, 1980, 590.

18. **Anon.,** Pectin, in *The United States Pharmacopeia,* 20th rev., Suppl. 3, United States Pharmacopeial Convention, Rockville, 1982, 565.

19. **Anon.,** Pectin, *The United States Pharmacopeia,* 20th rev., Suppl. 3, United States Pharmacopeial Convention, Rockville, 1983, 707.

20. **Cox, R. E. and Higby, R. H.,** A better way to determine the jellying power of pectins, *Food Ind.,* 16, 441, 1944.

21. Institute of Food Technologists, Pectin standardisation, final report of the IFT committee, *Food Technol.,* 13, 496, 1959.

22. **Steinhauser, J., Otterbach, G., and Gierschner, K.,** Vergleich von Metoden zur Bestimmung der Gelierkraft von Pektin, *Ind. Obst Gemüseverwert.,* 64, 179, 1979.

23. **Joseph, G. H. and Baier, W. E.,** Methods of determining the firmness and setting time of pectin test jellies, *Food Technol.,* 3, 18, 1949.

24. **Hinton, C. L.,** The setting temperature of pectin jellies, *J. Sci. Food Agric.,* 1, 300, 1950.

25. **Anon.,** Gel power of low ester pectin, in *Food Chemicals Codex,* 2nd ed., National Academy of Science, Washington, D.C., 1972, 580.

26. **Cambell, L. A. and Palmer, G. H.,** Pectin, in *Topics in Dietary Fiber Research,* Spiller, G. A., Ed., Plenum, New York, 1978, 105.

27. **Anon.,** Evaluation of certain food additives, 25th report of the Joint FAO/WHO Expert Committee on Food Additives, World Health Organization, Geneva, 1981.

28. **Anon.,** *Fed. Regist.,* 48, 51148, 1983.

29. **Anon.,** Departmental consolidation of the food and drugs act and of the food and drug regulations with amendments to August 5, 1982, Department of National Health and Welfare, Canada, 1981.

30. **Karpovich, N. S., Telichuk, L. K., Donchenko, L. V., and Totkajlo, M. A.,** Pectin and raw material resources, *Pishch. Promst. (Moscow),* 3, 36, 1981.

31. **Ishii, S., Kiho, K., Sugiyama, S., and Sugimoto, H.,** Low methoxyl pectin prepared by pectin-esterase from *Aspergillus japonicus, J. Food Sci.,* 44, 611, 1979.

32. **Kohn, R., Markovič, O., Machová, E.,** Deesterification mode of pectin by pectin esterases of *Aspergillus foetidus,* tomatoes and alfalfa, *Collect. Czech. Chem. Commun.,* 48, 790, 1983.

33. **Barret, A. J. and Northcote, D. H.,** Apple fruit pectic substances, *Biochem. J.,* 94, 617, 1965.

34. **Aspinall, G. O., Craig, J. W. T., and Whyte, J. L.,** Lemon peel pectin. I. Fractionation and partial hydrolysis of water-soluble pectin, *Carbohydr. Res.,* 7, 442, 1968.

35. **Rees, D. A. and Wight, A. W.,** Polysaccharide conformation. VII. Model building computation for α-1,4 galacturonan and the kinking function of L-rhamnose residues in pectic substances, *J. Chem. Soc. B,* 1366, 1971.

36. **Palmer, K. J. and Hartzog, M. B.,** An X-ray diffraction investigation of sodium pectate, *J. Am. Chem. Soc.,* 67, 2122, 1945.

37. **Walkinshaw, M. D. and Arnott, S.,** Conformations and interactions of pectins. I. X-ray diffraction analyses of sodium pectate in neutral and acidified forms, *J. Mol. Biol.,* 153, 1055, 1981.

38. **Walkinshaw, M. D. and Arnott, S.,** Conformations and interactions of pectins. II. Models for junction zones in pectinic acid and calcium pectate gels, *J. Mol. Biol.,* 153, 1075, 1981.

39. **Morris, E. R., Powell, D. A., Gidley, M. J., and Rees, D. A.,** Conformations and interactions of pectins. I. Polymorphism between gel and solid states of calcium polygalacturonate, *J. Mol. Biol.,* 155, 507, 1982.

40. **Powell, D. A., Morris, E. R., Gidley, M. J., and Rees, D. A.**, Conformations and interactions of pectins. II. Influence of residue sequence on chain association in calcium pectate gels, *J. Mol. Biol.*, 155, 517, 1982.
41. **Speiser, R., Copley, M. J., and Nutting, G. C.**, Effect of molecular association and charge distribution on the gelation of pectin, *J. Phys. Colloid Chem.*, 51, 117, 1947.
42. **Smit, C. J. B. and Bryant, E.**, Properties of pectin fractions separated on diethylaminoethyl-cellulose columns, *J. Food Sci.*, 32, 197, 1967.
43. **Heri, W., Neukom, H., and Deuel, H.**, Chromatographische Fraktionierung von Pektinstoffen in Diäthylaminoäthyl-cellulose, *Helv. Chim. Acta*, 44, 1939, 1961.
44. **Heri, W., Neukom, H., and Deuel, H.**, Chromatographic von Pektinen mit verschiedener Verteilung der Methylester-Gruppen auf den Fadenmolekülen, *Helv. Chim. Acta*, 44, 1945, 1961.
45. **van Deventer-Schriemer, W. H. and Pilnik, W.**, Fractionation of pectins in relation to their degree of esterification, *Lebensm. Wiss. Technol.*, 9, 42, 1976.
46. **Taylor, A. J.**, Intramolecular distribution of carboxyl groups in low methoxyl pectins — a review, *Carbohydr. Polym.*, 2, 9, 1982.
47. **Tuerena, C. E., Taylor, A. J., and Mitchell, J. R.**, Evaluation of a method for determining the free carboxyl group distribution in pectins, *Carbohydr. Polym.*, 2, 193, 1982.
48. **Baig, M. M., Burgin, C. W., and Cerda, J. J.**, Fractionation and study of chemistry of pectic polysaccharides, *J. Agric. Food Chem.*, 30, 768, 1982.
49. **Neukom, H., Amadò, R., and Pfister, M.**, Neuere Erkenntnisse auf dem Gebiete der Pektinstoffe, *Lebensm. Wiss. Technol.*, 13, 1, 1980.
50. **De Vries, J. A., Rombouts, F. M., Voragen, A. G. J., and Pilnik, W.**, Enzymatic degradation of apple pectins, *Carbohydr. Polym.*, 2, 25, 1982.
51. **De Vries, J. A., den Vijl, C. H., Voragen, A. G. J., Rombouts, F. M., and Pilnik, W.**, Structural features of the neutral sugar side chains of apple pectic substances, *Carbohydr. Polym.*, 3, 193, 1983.
52. **Owens, H. S., Lotzkar, H., Schultz, T. H., and Maclay, W. D.**, Shape and size of pectinic acid molecules deduced from viscometric measurements, *J. Am. Chem. Soc.*, 68, 1628, 1946.
53. **Christensen, P. E.**, Methods of grading pectin in relation to the molecular weight (intrinsic viscosity) of pectin, *Food Res.*, 19, 163, 1854.
54. **Sorochau, V. D., Dzizenko, A. K., Bodin, N. S., and Ovodov, Y. S.**, Light-scattering studies of pectic substances in aqueous solution, *Carbohydr. Res.*, 20, 243, 1971.
55. **Davis, M. A. F., Gidley, M. J., Morris, E. R., Powell, D. A., and Rees, D. A.**, Intermolecular association in pectin solutions, *Int. J. Biol. Macromol.*, 2, 330, 1980.
56. **Berth, G., Anger, H., Plashchina, I. G., Brando, E. E., and Tolstoguzov, V. B.**, Structural study of the solutions of acidic polysaccharides. II. Study of some thermodynamic properties of the dilute pectin solutions with different degrees of esterification, *Carbohydr. Polym.*, 2, 1, 1982.
57. **O'Beirne, D. and van Buren, J. P.**, Size distribution of high weight species in pectin fractions from Idared apples, *J. Food Sci.*, 48, 276, 1983.
58. **Padival, R. A., Ranganna, S., and Manjrekar, S. P.**, Stability of pectins during storage, *J. Food Technol.*, 16, 367, 1981.
59. **Michel, F., Doublier, J. L., and Thibault, J. F.**, Investigations on high-methoxyl pectins by potentiometry and viscometry, *Prog. Food Nutr. Sci.*, 6, 367, 1982.
60. **Rinaudo, M.**, Comparison between results obtained with hydroxylated polyacids and some theoretical models, in *Polyelectrolytes*, Selegny, E., Ed., Reidel, Dortrecht, 1974, 157.
61. **Ravanat, G. and Rinaudo, M.**, Investigation on oligo- and polygalacturonic acids by potentiometry and circular dischroism, *Biopolymers*, 19, 2209, 1980.
62. **Smit, C. J. B. and Bryant, E. F.**, Ester content and jelly pH influences on the grade of pectins, *J. Food Sci.*, 33, 262, 1968.
63. **Kohn, R.**, Ion binding on polyuronates — alginate and pectin, *Pure Appl. Chem.*, 42, 371, 1975.
64. **Scott, J. E.**, Fractionation by precipitation with quaternary ammonium salts, *Methods Carbohydr. Chem.*, 5, 38, 1965.
65. **Stutz, E. and Deuel, H.**, Polyampholyte mit verschiedener Ladungsverteilung, *Helv. Chim. Acta*, 38, 1757, 1955.
66. **Glahn, P. E.**, Hydrocolloid stabilization of protein suspensions at low pH, *Prog. Food Nutr. Sci.*, 6, 171, 1982.
67. **Neukom, H. and Deuel, H.**, Über den Abbau von Pektinstoffen bei alkalischer Reaktion, *Z. Schweiz. Forstv.*, 30, 223, 1958.
68. **Morris, E. R., Gidley, M. J., Murray, E. J., Powell, D. A., and Rees, D. A.**, Characterization of pectin gelation under conditions of low water activity, by circular dichroism, competitive inhibition and mechanical properties, *Int. J. Biol. Macromol.*, 2, 327, 1980.
69. **Thom, D., Dea, I. C. M., Morris, E. R., and Powell, D. A.**, Interchain associations of alginate and pectins, *Prog. Food Nutr. Sci.*, 6, 97, 1982.

70. **Mitchell, J. R.,** Rheology of gels, *J. Texture Stud.,* 7, 313, 1976.
71. **Kohn, R. and Luknar, O.,** Intermolecular calcium ion binding on polyuronates — polygalacturonate and polyguluronate, *Collect. Czech. Chem. Commun.,* 42, 731, 1977.
72. **Bystricky, S., Kohn, R., and Sticzay, T.,** Effect of polymerization degree of oligogalacturonates and D-galacturonans on their circular dichroic spectra, *Collect. Czech. Chem. Commun.,* 44, 167, 1979.
73. **Rees, D. A.,** Polysaccharide gels, *Chem. Ind. (London),* 630, 1972.
74. **Rees, D. A.,** Polysaccharide conformation in solutions and gels — recent results on pectins, *Carbohydr. Polym.,* 2, 254, 1982.
75. **Anon.,** GENU handbook for the fruit processing industry, The Copenhagen Pectin Factory Ltd., Lille Skensved (Denmark), 1984.
76. **Anon.,** Fruit preparations for yoghurt, The Copenhagen Pectin Factory Ltd., Lille Skensved (Denmark), 1980.
77. **Anon.,** Confectionery products with GENU pectins, The Copenhagen Pectin Factory Ltd., Lille Skensved (Denmark), 1981.
78. **Anon.,** Stabilization of fermented and directly acidified sour milk drinks, The Copenhagen Pectin Factory Ltd., Lille Skensved (Denmark), 1982.
79. **Pfeifer and Langen KG.,** U.S. Patent 3,595,676, 1965.
80. **Valet, R.,** Zur Herstellung von Fruchtzubereitungen und deren Anwendung in Milchprodukten, *Ind. Obst. Gemüseverwert.,* 67, 507, 1982.
81. **Termote, F., Rombouts, F. M., and Pilnik, W.,** Stabilization of cloud in pectinesterase active orange juice by pectic acid hydrolysates, *J. Food Biochem.* 1, 15, 1977.
82. **Röcken, W.,** Die Bedeutung von Pektin und Pektinasen für die Trubstabilität von Orangelimonaden, *Brauwelt,* 8, 224, 1979.
83. **Schopf, L. D., Sakowicz, J. K., and Trenk, H. L.,** U.S. Patent 4,321,279, 1980.
84. **Christensen, S. H.,** Pectin a natural hydrocolloid for confectionery products, *Confect. Prod.,* 43, 378, 1977.
85. **Barwick, B. E. and Sneath, M. E.,** U.S. Patent 4,119,739, 1978.
86. **Exler, H.,** German Patent 270,938, 1969.
87. **Arolski, A. T., Usheva, V. B., Gruev, P. V., Richev, G. T., and Doucheva, Z. S.,** U.S. Patent 4,031,264, 1977.
88. **Nesmeyanov, A. N., Rogozhin, S. V., Tolstoguzov, V. B., Misjurev, V. I., Erchova, V. A., and Braudo, E. E.,** British Patent 1,474,891, 1977.
89. **Buckley, K. and Mitchell, J. R.,** U.S. Patent 3,973,051, 1976.
90. **Waitman, R. H. and Hoos, J. W.,** U.S. Patent 3,367,784, 1978.
91. **Toft, K.,** Interactions between pectins and alginates, *Prog. Food Nutr. Sci.,* 6, 89, 1982.
92. **Jamison, J. D., Towle, G. A., and Vermeychuk, J. G.,** U.S. Patent 4,129,663, 1978.
93. **Nelson, F. F.,** Newer applications for pectin, *Food Prod. Dev.,* 13, 38, 1979.
94. **Leo, H. T. and Taylor, C. C.,** U.S. Patent 2,754,214, 1956.
95. Sanei Chemical Industries, Japanese Patent 53,124,661, 1978.
96. **Kratschanov, C., Stancov, S., Popova, M., and Pancheva, T.,** Anwendung von Pektinemulgatoren zur Herstellung von Lebensmittelemulsionen mit reduziertem Energiewert, *Nahrung,* 26, 217,1982.
97. **Tokunaga, K., Okuyama, G., Nagasawa, H., and Otani, Y.,** Effects of pectin on formation and stability of emulsions, *Cosmet. Toiletries,* 96, 30, 1981.
98. **Wegener, J. B., Baer, B. H., and Rodgers, P. D.,** Improving quality of frozen strawberries with added colloids, *Food Technol.,* 5, 76, 1951.
99. **Swenson, H. A., Miers, J. C., Schultz, T. H., and Owens, H. S.,** Pectinate and pectate coatings. II. Application to nuts and fruit products, *Food Technol.,* 7, 232, 1953.
100. **Gurkin, M., Sanderson, G. W., and Graham, H. N.,** U.S. Patent 3,666,484, 1972.
101. **Mylaeus, A.,** British Patent 1,295,007, 1972.
102. Moldavian Food Industry Research Institute, U.S.S.R. Patent 542,505, 1977.

Index

INDEX

T

U